DEVELOPMENTS IN PETROLEUM GEOLOGY—1

DEVELOPMENTS IN PETROLEUM GEOLOGY—1

Edited by

G. D. HOBSON

Consultant, V. C. Illing & Partners,
Cheam, Surrey, UK

APPLIED SCIENCE PUBLISHERS LTD
LONDON

APPLIED SCIENCE PUBLISHERS LTD
RIPPLE ROAD, BARKING, ESSEX, ENGLAND

ISBN: 0 85334 745 X

WITH 106 ILLUSTRATIONS

Printed in Great Britain by Galliard (Printers) Ltd, Great Yarmouth

PREFACE

The past twenty-five years have seen a great expansion in knowledge relating to petroleum geology. This is not a case of developing new basic concepts, but rather a matter of having obtained new and better supporting data, and the acquisition of information on significant details by the use of new techniques and instruments.

Those active in particular branches of petroleum geology will have kept pace with developments relating to such branches, but it would require a most voracious reader or well-financed conference attender to cover fully the many aspects of this subject. Indeed, it might be wondered whether such a person, if he or she exists, could be doing much beyond acting as a sponge for information. The contents of the *American Association of Petroleum Geologists Bulletin* were equivalent to about 1·2 million words in the 1952 volume and 1·9 million words in 1976, having gone through a peak of some 2·2 million words in 1974. This is by no means the only journal containing papers concerned with points of interest in petroleum geology.

The basic aim of this volume is to put in the picture those who are not already experts in a particular branch of petroleum geology so that they may become aware of what has been going on elsewhere during recent years and to enable them, where applicable, to make use of such knowledge in their own work. Various considerations have precluded the writing of exhaustive texts, including the fact that many readers might find them overwhelming; yet it is believed that there has been judicious selection of detail, and that the bibliographies will allow the ready acquisition of additional information where required.

Team-work is an essential feature in the effective search for and development of oil and gas fields, for the application of petroleum geology does not end with the drilling of a discovery well. Hence, an understanding of the techniques and the kinds of information employed by other members of the team can be beneficial. It is also important to recognise the limitations of techniques and the uncertainties which can be inherent in the interpretation of observations. Sometimes over-enthusiasm may lead to the inappropriate application of some techniques or ideas, and to a failure to recognise the possibility of alternative explanations of the observations. Cautionary notes are to be found in various places in this volume.

The fascinating subject of plate tectonics, the modern and decidedly more all-embracing version of continental drift, bids fair to co-ordinate much in geology that tended formerly to be treated separately. The patterns of occurrence of mineral deposits have begun to receive consideration against the background of plate tectonics. Major tilts, fractures and displacements affecting lithospheric plates may lead to the advance or retreat of seas as sedimentary basins evolve, and there are effects which determine the nature, extent, mode of emplacement and thickness of sedimentary sequences. Temperature levels play a major role in the development and evolution of oil and natural gas, and variations in temperature history from place to place are dependent on factors involved in plate tectonics.

Detailed knowledge on the components in crude oil and in solvent extracts from sediments has grown apace, while ideas are developing on the structure of the intractable solid organic matter and on the origin of the asphaltic materials in sediments. Although details concerning the formation of petroleum have become much clearer, important aspects of the mechanism of oil and gas migration remain to be elucidated. Water and its movement and associations in sediments are of importance in various connections: migration, mineralogical changes and mineral redistribution which affect porosity and permeability, and even general geological structure; while what may be considered to be less than normal water displacement leads to over-pressured clays, with bulk densities lower than are usual for their depths of burial.

Wells provide the only direct data on some matters such as lithology, fluid content of the rocks, etc., and in certain cases sound deductions concerning the origins and alterations of the rocks may be made with considerable assurance from observations made in wells or on cores, cuttings and fluids obtained from wells. In view of the high cost of wells, as drilling depths have increased and oil exploration has moved offshore, there has been a strengthening of the incentive to make deductions about the fluid content or the nature of the rocks in advance of drilling. Seismic surveys have long been used to try to ascertain rock boundary shapes and arrangements suitable for trapping hydrocarbons. Improvements in the acquisition and treatment of seismic data have aided in the better definition of possible trapping situations. At the same time features have been recognised on seismic records which could be indicative of the presence of free gas in the rocks. In addition, the prospect of satisfactorily deducing something about the lithology from seismic observations has emerged. Certain characteristics of the processed seismic data are involved, as well as the shapes which they indicate. In the latter respect it is clear that reflection seismic surveys can provide information at laterally far closer spacing than is ever practicable by drilling deep wells.

The contributors to this volume worked with an awareness of the general aims and a knowledge of the proposed titles of other chapters; they did not have access to the manuscripts of the other chapters. Their efforts, including the speed with which they dealt with minor queries arising from the original texts, are gratefully acknowledged.

CONTENTS

LIST OF CONTRIBUTORS

BRIAN S. COOPER

Robertson Research International, Ty'n-y-Coed, Llanrhos, Llandudno LL30 1SA, UK.

WALTER H. FERTL

Director of Interpretation and Field Development, Dresser Atlas Division, Dresser Industries Inc., PO Box 1407, Houston, Texas 77001, USA.

KINJI MAGARA

Imperial Oil Ltd, 500 Sixth Avenue South West, Calgary, Alberta T2P OS1, Canada.

MILES F. OSMASTON

Consultant Engineer, The White Cottage, Sendmarsh, Ripley, Woking, Surrey GU23 6JT, UK.

J. R. PARKER

Shell U.K. Exploration and Production, Shell Centre, London SE1 7NA, UK.

R. C. SELLEY

Department of Geology, Royal School of Mines, Imperial College, Prince Consort Road, London SW7, UK.

R. E. SHERIFF

Seiscom Delta Inc., PO Box 36789, *Houston, Texas* 77036, *USA.*

CHARLES B. STONE

President, Stone Geophysical Company, PO Box 1778, *Conroe, Texas* 77301, *USA.*

J. C. M. TAYLOR

V. C. Illing & Partners, 'Cuddington Croft', Ewell Road, Cheam, Sutton, Surrey SM2 7NJ, UK.

B. TISSOT

Institut Français du Pétrole, 4 *avenue de Bois-Préau,* 92502 *Rueil-Malmaison, France.*

Chapter 1

SOME FUNDAMENTAL ASPECTS OF PLATE TECTONICS BEARING ON HYDROCARBON LOCATION

MILES F. OSMASTON

Ripley, Surrey, UK

SUMMARY

The interpretive framework of plate tectonics is carefully reassessed and considerably extended with particular regard to the problems of hydrocarbon exploration. It is found that the seismic low-velocity zone lies wholly within the plates, which are thus very much thicker and stiffer than previously recognised. The consequences impinge upon surface and near-surface phenomena to a quite remarkable degree, offering (inter alia) *major and unifying advances in the interpretation of basins, whether in orogenic settings, at ocean margins, or within continents. Specifically, the persistent and episodic differential epeirogenic movement of block-and-basin crustal mosaics is shown to be characteristic of basin complexes formed by limited plate separation at a much earlier time (even as long ago as Late Precambrian). Island inliers (micro-continents) detached at that time provide buried structural highs important for hydrocarbon accumulation. Extremely precise (∼5 km) reconstructions are achievable.*

Plate stiffness makes epeirogenic processes at plate edges felt at great distances from them, often causing flexural failure, intra-plate rifting, intra-plate volcanism and further epeirogeny. Thick plates enhance thermal epeirogenic effects, enable shelf emplacement of ophiolite slices to occur during early separation, explain certain features of the subduction process, and shed light on the incidence and petrogenetic mechanisms of intra-plate volcanism. Correct treatment of plate thermal epeirogeny shows the previous wholly-vertical contraction/expansion requirement to be erroneous.

1

The plate tectonics of the Phanerozoic of north-western Europe and the post-Palaeozoic of the Middle East are among the examples discussed.

With this greatly increased relevance and precision, plate tectonic analysis is clearly capable of playing a valuable and detailed part in hydrocarbon exploration, contributing directly both to the precise delineation of basement structural outlines and to the interpretation of subsequent structural, sedimentary and thermal evolution.

INTRODUCTION

The location of hydrocarbon accumulations requires the detailed application of a highly developed technical and interpretive expertise. At present, however, the role of plate tectonics in this process is rarely more than to provide a fascinating backdrop whose detail is nowhere precise enough for direct use. Margins appear blurred, the plate motion significance of particular events obscure or ambiguous and, even when the causative motion is clear, the geological consequences sometimes vary sharply and apparently unpredictably from place to place. Are these merely the teething problems of a new scientific field or are they truly random features bringing fundamental limitations upon its economic usefulness? In particular, can plate tectonics be made sufficiently precise and comprehensive to contribute usefully to the detailed interpretation (and prediction?) of the structural, sedimentary and thermal development of sedimentary basins?

The high costs of exploration make these questions extremely important. This chapter shows that certain widely accepted features of plate tectonics are capable of radical reappraisal, and the effect of this will be shown not only to widen greatly the probable applicability and scope of interpretations based on plate tectonics, but also to offer a possible way of making them more precise and economically useful. To do so requires an excursion into deep geophysical matters far beyond the shallow effects we seek to interpret.

Plate tectonics as a field of study rests on the firm establishment of two essential discoveries. One is that the Earth's lithosphere† is formed of a rather small number of major plates which, to a remarkable degree, behave as mechanical entities and are outlined by belts where they interact.[1,2] The

† The term is used here in its usual plate tectonic sense to include both the crust and a substantial thickness of mantle material to a depth beyond which flow-creep becomes the dominant feature of any response to stresses imposed at rates typical of plate motion.

other is that this interaction consists of large-scale separative motion involving lithosphere creation, and commensurate approximative‡ motion made possible by lithosphere consumption or plate-edge deformation.[3,4]

A thorough understanding of the basic processes involved in large-scale lithosphere genesis and consumption is a central aspect of plate tectonics study, as the way to enable an interpreter to relate the general to detailed matters of exploration concern.

The grand scale of plate genesis provides widespread clues about the basic mechanisms of oceanic plate evolution. In particular, the coherence of the data leaves no doubt that the (3 km +) subsidence of ocean floors with increasing age is due to progressive heat loss from and density increase *within the plate itself*, thus stressing the direct role of plate tectonics in long-term epeirogenic movement and suggesting that sub-plate processes (e.g. convection) may have, for practical purposes, no direct surface manifestation apart from the plate motions themselves. Evidence of the processes involved in subduction, on the other hand, although supporting the general picture of calc–alkaline volcanism induced by shearing-produced heat at Benioff planes, seems in detail to be much less coherent than ocean floor data, and this may partly be due to the fact that here we have the interaction of two plates, each of which has a distinctive prior history and resulting constitution.

The highly ordered features of large parts of the ocean floor seem a far cry from the complex plate fragmentation and reorganisation which must have characterised plate interaction elsewhere (e.g. the Alpide belt, the east Asian margin, or western Europe), if the resulting mosaics of basins and highs are to prove tractable in plate tectonics terms. However, it seems wholly premature to invoke some entirely different plate interaction process for these until all the possibilities for complex interplay of 'normal' kinds of plate genesis and consumption have been fully explored. In regions such as these it is important to remember that interactions between major plates probably proceed at typical plate relative speeds (5 cm/y is 250 km in 5 my), so that what now appears to have been a continuum of interaction may turn out to have been a series of rapid discrete phases blurred by overlapping after-effects. As discussed later, there seems to be evidence for this in the Caledonian–Appalachian orogenic belt. The notion that plate motions do

‡ As the kinematical opposite of *separative* this term is preferable to the frequently used *convergent*, which is ambiguous; having also a static geometrical meaning, to which use in plate tectonics it should desirably be restricted. Otherwise, consider the possible confusion when discussing the convergence (motion) of convergent (geometrical) margins.

not start or stop abruptly was born of unrestrained convection hypotheses but is probably incorrect, especially in situations where motions may be guided or limited by plate obstructions.

For the basin analyst a crucial question concerns the duration and nature of the after-effects of plate interaction, particularly separation. Indeed, its initial main concern with the nature of interaction at plate boundaries made some geologists feel that plate tectonics could have little bearing on geological events at places that were far from a plate boundary at the time. Recent developments make it clear, however, that the after-effects of lithosphere genesis are active for much longer and are far more widespread than seemed likely a few years ago. Because these after-effects are mainly thermal, epeirogenic and sedimentary in character, and provide important control on the location of faults, they bear on many aspects of hydrocarbon occurrence.

We discuss first a wide variety of evidence bearing upon the thickness and constitutional variation of plates, showing how it is upon these that much of the residual (and sometimes highly correlated) activity at the surface depends. This provides a basis for the treatment of basins (*chasmic* basins) of true plate separation origin and yields strong arguments that such basins are also a widespread feature of the continents. Attention will be drawn both to the influences of various tectonic environments upon the potentially lengthy evolutionary course of chasmic basins, and to the peripheral and other effects associated with the separative mode of origin. Finally, we will consider what happens when plates thus constituted are involved in approximative motions, both subduction and collision. There are numerous factual examples to guide us in the construction of this interpretive framework.

PLATE THICKNESS AND ITS IMPLICATIONS

Plate thickness is a matter of direct significance for the evolution of basins of plate separative origin because, as we shall see, the amount and especially the duration of the ensuing subsidence are greater when the plate thus generated is very thick. A related effect is that the heat introduced into a thick plate, when magma intrudes it from its base, is greater and produces larger and more extensive epeirogenic movements at the surface. Thickness also affects a plate's flexural stiffness, and hence the lateral distance to which plate epeirogenic behaviour can be influenced by density anomalies which develop within it.

Before the days of plate tectonics, adherents of continental drift thought that perhaps only the continental crust was involved in the motion, but the currently most widely held view, that plate thickness under oceans is 50–100 km, is based on a remarkable coincidence of information from four different kinds of study. It is now likely, however, that this coincidence has another explanation and that plate thicknesses are very much greater even than this. The reasons are outlined in the following sections.

Existing Ideas on Plate Thickness

When the seismic shear wave low velocity (LV) zone, which has a well defined top at an average depth of about 70 km beneath ocean basins, came to be interpreted as a zone of partial melting,[5] it was quickly adopted as marking the base of tectonic plates. Petrogenetic studies,[6-8] suggesting that magmas like the bulk of those found on oceanic islands underwent primary segregation from mantle material at depths of 60–80 km, apparently supported this view. Another indication came from theoretical studies of plate genesis by the continuous lateral accretion of hot material at the axes of mid-ocean ridges. These suggested that the heat embodied in the accreted plate material is gradually dissipated upwards through the ocean floor, resulting in a gradual density increase and subsidence which is evident as the sloping flanks of ocean ridges. Analyses of age, heat flow and subsidence data from the ocean floor have consistently shown that the main feature is the progressive cooling of a slab of material 50–100 km thick.[9-14] Finally, there was the discovery that seismic wave travel times in the neighbourhood of Benioff zones outline a tongue of relatively cool descending lithosphere, which is also of about this thickness initially.[15-17]

Can all these apparent indications of plate thickness really be mistaken?

The Thick-Plate Model

Notice first that there could be a substantial difference between lithosphere thickness and plate thickness, for the former relates to material in a particular physical state whereas the latter is concerned with the entire thickness of co-moving material; the two would be similar only if the underside of the lithosphere were vigorously swept by differently moving material. It is not clear whether this is anywhere the case.

Evidence that at least the continental parts of plates may be very thick is of long standing. In 1961 Bernal[18] pointed to the fact that some Benioff zones extend to nearly 700 km depth as an indication that the LV zone might merely constitute 'the paste in a sandwich', the whole of which could undergo rifting. Since 1960, evidence has been accumulating that the geophysical properties beneath continents and oceans differ to a depth of at least 400 km,[19−25] and there is evidence from volcanism that they are geochemically distinct to as much as 400 km depth also;[26] all of which implies that, beneath continents at least, the material to this depth retains more or less permanently its relation to the continent above. Notice particularly that this depth comfortably encompasses the entire sub-continental LV zone, where present. Recently, Morgan[27] was prepared to envisage an oceanic plate thickness of 150 km, thus including a substantial thickness of what he considered to be asthenospheric material.

The main question concerns the physical significance of the LV zone. In fact, a fluid content of only 0·1 % to 1 % could be enough to produce the observed seismic velocity and attenuation properties, if the fluid were present in the likely form of very thin intergranular films[28,29] instead of the globules previously envisaged by Birch.[30] Thus, it could be scarcely more than the presence of free volatiles,[31] and not a substantial melt, that is being observed. This type of interpretation is strongly favoured by other considerations. The LV zone is not equally present everywhere, but is well marked only where the heat flow being transmitted upwards through the relevant depth range probably exceeds a value in the region of 30 mW/m² (0·7 HFU); this includes the oceans and active orogenic belts but excludes many shield areas (where total heat flow, including the crustal contribution, is sometimes no more than this). Recent work in the Pacific[32,33] shows that, as cooling proceeds, the LV zone deepens to about 55 km during the first 30 my and then more slowly to 85 km at 100 my. Furthermore, the seismic velocities in the 'lid' above the LV zone have been shown[30] to imply a temperature gradient which attains near-solidus temperatures at the LV zone.

A point which has been overlooked hitherto is that any build-up of interstitial fluid would produce a marked lowering of the thermal conductivity, especially if any of the volatile content were present as a gas phase, which Eggler[34] has shown is likely in the case of CO_2. As the LV layer is 100–250 km thick it is therefore probable that the thermal resistance of the layer largely controls the heat flow through it. In that case the influence of its fluid content upon its thermal resistance will have a strongly stabilising effect on the heat flow; any cooling within the layer would (since the temperature of its top is, by definition, close to the solidus) reduce the fluid content in the layer, and the resulting decrease in overall thermal resistance would tend to prevent the heat flow from dropping. Ocean floor heat flow and subsidence studies have suggested that the heat flow out of the top of the oceanic LV zone is indeed remarkably constant, but this was regarded by McKenzie[13] as strong evidence for heat transport by convective flow to near the top of the LV zone. It appears, however, that an LV zone entirely integral with the plate would have the same property.

It also follows from this argument that within the LV zone the geotherm will tend to follow a line giving constant fluid content (or to be more exact, constant thermal conductivity), which would explain why, in any one place, the seismic shear velocity, V_s, rises only slowly throughout the thickness of the LV layer. As the plate ages, cooling will eventually lower the level of fluid content in the layer and raise V_s as is seen beneath parts of continents. The base of the LV layer is probably determined by the rise in (solid) thermal conductivity at depth, which will flatten the temperature gradient. The layer will finally disappear when the slope of the geotherm is too low to produce interstitial fluid at any depth. A possible sequence of geotherms is sketched in Fig. 1.

The proposed plate model which emerges has a thickness which extends at least to the *base* of the LV zone (where present) at a depth which ranges to more than 300 km. The LV zone is due to interstitial free volatiles and incipient melting, giving a fluid content probably ranging up to rather more than 1 % and stabilising the transmitted heat flow as outlined above. If sedimentary rocks are any guide, such a low fluid content would have a negligible effect upon the structural integrity of the material. Its temperature, on the other hand, almost certainly implies susceptibility to creep. Nevertheless, the conclusion we shall reach in succeeding sections is that not only does this material not, in fact, undergo general displacement with respect to the material above, but it contributes substantially to the overall plate stiffness, enabling plates to transmit flexural stresses for long times (several tens of my) over large distances (upwards of 1000 km).

These conclusions make very little difference to the interpretation of the subsidence pattern and sedimentary evolution of plate areas generated within the past 80 my, because this is the interval during which the cooling of plate material above the LV layer is the dominant factor. They are likewise consistent with the seismically observed cool tongue effect at Benioff zones, for the deeper material in the plate will have cooled too little to be seismically distinguishable. They are immensely significant, however, for older basins of plate separative origin because it means that these will continue to subside and to constitute receptacles for sediments throughout the greatly increased time scale required for the slow cooling of the LV layer and any even deeper material in the plate. Recent ocean floor data confirm the first part of this slow process, in the form of markedly slower subsidence (0·5 km total) between 80 my and 140 my.[14,32,35] Our result is relevant

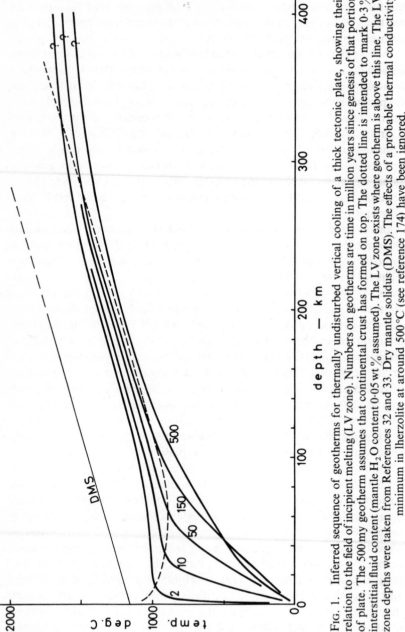

Fig. 1. Inferred sequence of geotherms for thermally undisturbed vertical cooling of a thick tectonic plate, showing their relation to the field of incipient melting (LV zone). Numbers on geotherms are time in million years since genesis of that portion of plate. The 500 my geotherm assumes that continental crust has formed on top. The dotted line is intended to mark 0·3 % interstitial fluid content (mantle H_2O content 0·05 wt % assumed). The LV zone exists where geotherm is above this line. The LV zone depths were taken from References 32 and 33. Dry mantle solidus (DMS). The effects of a probable thermal conductivity minimum in lherzolite at around 500 °C (see reference 174) have been ignored.

therefore, to the proposal[36] that many long-subsiding continental basins owe their existence to a corresponding amount of plate separation at a much earlier time. We shall return to this later.

Intra-Plate Volcanism in the Context of Thick Plates

Intra-plate volcanism or intrusion can have a crucial impact upon the prospectivity of an area, so an understanding of what controls its occurrence is very desirable. The fact that intra-plate volcanism occurs at all, and often so far from zones of plate interaction, has seemed to many to require a cause which was separate and additional to the basic concepts of plate tectonics. In this respect the hypothesised existence of penetrative plumes from the lower mantle[37] has already provided a popular scapegoat for innumerable volcanic occurrences, completely supplanting the LV zone as the presumed local source of the corresponding melts. There are, indeed, major objections to the plume hypothesis, especially to its application in continental areas.[26,38] Others have invoked abortive tectonic rifting as a control. The concept of thick plates offers a new approach to this problem.

The formation and slight opening of vertical cracks through plates, due to plate bending or cooling shrinkage, for example, will draw material into the crack from the base of the plate. Because the plate base is inevitably a viscosity gradation, the earliest material to enter the crack will have lain within the lower part of the plate. If the vertical extent of the crack is sufficient, the material rising in it will undergo an increasing degree of partial melting. The resulting reduction in density and effective viscosity will endow the column of material with increasing ability to force its way upward. When a level is reached at which the temperature contrast between the fracture walls and the rising material is sufficient, the rapid heat loss to the walls will make the unmelted pieces grow and jam in the crack or freeze to the walls leaving the segregated magma to continue its ascent.

Now the low (but still super-adiabatic) thermal gradient, which we have inferred to characterise the seismic LV layer, changes to a much steeper gradient not far above (Fig.1). Consequently, any partially melted material rising in a fracture will experience a steep decrease in wall temperature on emerging from the layer, so magma segregation will set in strongly during the next few kilometres of ascent. Thus the fact that, as mentioned earlier, ocean island magmas do seem to have been segregated at 60–70 km depth is apparently quite consistent with the thick-plate model in so far as oceanic areas are concerned. O'Hara[39,40] has argued on petrogenetic grounds that removal of phenocrysts by wall accretion plays an essential role in the evolution of magmas but has not applied it to primary magmatic segregation. Our model explains why segregation mainly occurs at particular levels and produces distinctive magmas. Clearly this approach to magma genesis can ultimately lead to an estimate of plate thickness, on the basis that the material has to rise far enough in the crack to attain the requisite degree of melting before it reaches the segregation level.

Turning briefly to continental intra-plate volcanism, there is much evidence, partly of a petrogenetic nature, but mainly from the petrology of nodules, that some magmas in this class (e.g. kimberlites) undergo rapid transport from depths of 200–300 km.[41–43] Presumably, where the LV zone is weak or absent, the temperature gradient is nowhere steep enough to give a preferred level at which the high-melting-point and unmelted constituents are frozen out. The high and variable $^{87}Sr/^{86}Sr$ ratios in these magmas led Brooks *et al.*[26] to infer derivation from material which had long been part of the plate, which fits the proposed mode of origin rather well.

In summary, thick plates account for several features of intra-plate volcanism and render volcanic melts more available, not less. Cracking of the plate is all that is required. To develop the appropriate cracking stresses five principal mechanisms, which may act separately or in combination, are envisaged, namely

(a) horizontal thermal contraction of the plate at depth,
(b) tension due to horizontal thermal expansion induced elsewhere in the same plate,
(c) differential isostatic adjustment within a heterogeneous plate, following the differential density changes and shrinkage induced by long-term cooling,
(d) wrench faulting of non-straight boundaries, and
(e) structural failure of the plate when its margin is uplifted as a result of plate interaction.

Cracks which open too slowly or too little at depth will induce intrusion rather than explicit volcanism.

Genesis of Thick Plates at Mid-Ocean Ridges

If the foregoing thick-plate model is to be tenable, the young edges of plates must either be already very thick when generated or increase their effective thickness much more quickly than has been thought. In either case this is, in principle, a matter of cooling the mantle material, freshly emplaced under mid-ocean ridges, sufficiently to lower its fluid content to a value at which the material behaves as a structurally integral part of the plate. However, if, as seems likely, the amount of melt present in the freshly emplaced material is only a few percent after the crust-forming constituent has left it, effective structural integrity may be achieved straight away, or after only very slight cooling. Probably, therefore, thick plates are constructed as such at a nearly-vertical sided axial intrusion zone in which wall accretion is the dominant process.

Thermal Epeirogeny Due to Intra-Plate Volcanism and Early Plate Separation

The fact that probably the entire (> 3 km) epeirogenic range of the ocean floor, following its generation at an ocean ridge, is attributable to changes in the heat content of the underlying plate material, justifies serious consideration of the possibility that all epeirogenic movement not due to surface loading changes (erosion, sedimentation, sea-level changes, volcanic effusion, etc.) has a similar origin.[36]

Intra-plate volcanism and early plate separation both involve raising material from the region of the plate base and emplacing most or all of it between lithosphere that is substantially older and cooler. This has important epeirogenic implications, several aspects of which have been discussed previously.[36,44] The thicker the plate is, the hotter the source material and the more of it that is needed to fill a given amount of crack opening or plate separation. Consequently, a doubling (say) of inferred plate thickness may nearly treble the amount of excess heat available to be conducted laterally from the new material into the adjacent lithosphere. Magmas in transit to the surface will further supplement this heat. The resulting spread of heat was called *lateral heat flush* by Osmaston.[36] The heat will spread sideways and upwards, flowing down the resultant thermal gradient produced by superimposing the horizontal gradient generated by the emplacement zone upon the pre-existing vertical gradient in the original lithosphere. Where the original lithosphere involved is well cooled and has a low vertical gradient, the lateral heat flush (from a given volume of newly emplaced material) will be more marked and, in the course of time, will affect the lithosphere temperatures to greater horizontal distances. The horizontal extent of the lateral heat flush thermal anomaly is also influenced by how far down the sheet or column of newly emplaced material the lateral heat flush originates; the deeper it originates, the further it will spread horizontally while it ascends, and the longer it will take for the lateral heat flush to be dissipated. Thus thick plates enhance these aspects also. Finally, the relatively low thermal conductivity of continental crust means that the heat will tend to propagate further beneath it before it can be dissipated upwards.

The ways in which the spreading heat will lower densities and produce plate uplift have been treated elsewhere.[36,44] Suffice it to say here that (for example) a temperature rise of 400 °C at the base of the continental crust, tapering linearly to zero at the surface and at 200 km depth, would produce a minimum of 1·5 km of isostatically adjusted uplift.[44] Any erosion would, of course, tend to be offset by further isostatic response.

In the light of these arguments, the epeirogenic associations of intra-plate volcanism, well marked as doming in shield areas,[45] appear likely to have a lateral heat flush origin. If plates were not very thick the heat supply would almost certainly be too small for such an interpretation. Plate separation, on the other hand, clearly provides an enormously greater excess heat resource. This suggests that lateral heat flush from a separation margin might eventually propagate, and reduce densities in the plate, to horizontal distances comparable with the plate thickness. Upwarping would probably begin too quickly to distinguish between this and the more popular belief (based on the Cloos model) that doming precedes, rather than follows, the initiation of plate separation. Caution is necessary, however, because of the extremely slow rate at which heat propagates by conduction, even though aided by radiative conduction at depths beyond 100 km. Simple calculations suggest that the 'feather edge' of the heat flush pattern may reach little further than 100 km from the source in 50 my, and 170 km in 150 my. Consequently, where the epeirogenic effects extend much further than this, other factors must be present.

There are, in fact, two kinds of situation that seem to involve lateral heat flush. In one of these, typified by the more than 700 km wide post-Eocene upwarp of the Arabian Red Sea margin, the large scale of the tilting must be due to buoyancy induced near the Red Sea combined with a considerable measure of plate flexural rigidity.

In the other, typified by the British Isles, whose Caledonian and Hercynian structure has now been exhumed to a distance of 600 km from the edge of the continental shelf (Fig. 2), it is likewise inconceivable that the lateral heat flush could have spread more than (say) a third of this distance since separative movements began in this part of the North Atlantic. In this case, however, plate rigidity certainly cannot be invoked because of the extensive independent vertical movements which have occurred in the Irish Sea and Hebrides areas.[46–49] The disturbance in the Irish Sea lies in the same line as the north–south belt of early Tertiary volcanism which extends from north-west Scotland to the Bristol Channel (Fig. 2) and reached a peak within a few million years of the (60 my bp) start of a new phase of separation beyond Rockall Plateau, 500–1100 km to the west.[49,50] The author believes that the heat-flush-induced buoyancy near the new plate-edge generated sufficient bending moment to fracture the plate along the general line of a pre-existing weakness (seen as a chain of post-Devonian basins), causing this volcanism and renewed faulting. The intensity of the dyke swarms suggests a considerable heat flush input, which is inferred to have caused the extension of the region of exhumation well across England.

Almost the same thing appears to have occurred early and late in the Jurassic.[51] On these occasions the line of weakness was the now-buried Viking and Central Grabens of the North Sea (Fig. 2), which were likewise some 650–1000 km from the probable new lines of separation (Hatton–Rockall basin (?), Rockall Trough), and experienced major thermal events and periods of exhumation.

Across the world, a similar situation is seen in South Australia, where the Spencer Gulf–Flinders Range differential movements[52,53] and present seismicity[54] seem related to the strong Cenozoic uplift[53] of the Tasman Sea margin (1000–1500 km to the east) following the opening of the Tasman Sea.[55] Other examples appear to be the late Triassic basins of eastern North America, the more recent faulting and present seismicity along the Mississippi Embayment–St Lawrence River line[54] (750–1000 km from the Atlantic shelf edge), and the late Jurassic outbreaks of intense basaltic volcanism along the Paraná–Parnaiba axis of Brazil[56] (750–1000 km from the contemporaneously upwarped eastern margin). Rifting and volcanic events in Africa, too, may have a similar cause. In southern Africa, four post-Jurassic marginal regressions apparently correlate with volcanic outbursts in the interior.[57]

We shall return to these matters later, on account of the importance of the major *differential* vertical movements which commonly occur in lateral heat flush regimes. Interpretations based on sub-plate heat (convection, plumes, etc.), as an alternative to lateral heat flush, prejudice in advance any hope of explaining the detailed distribution of such movements.

In this section we have inferred that lateral heat flush, insignificant unless plates are thick, produces buoyancy effects which are structurally significant over large distances and provide our first evidence of the flexural stiffness of plates. In the next section we look at another source of such evidence which has profound implications for the behaviour of 'passive' continental margins.

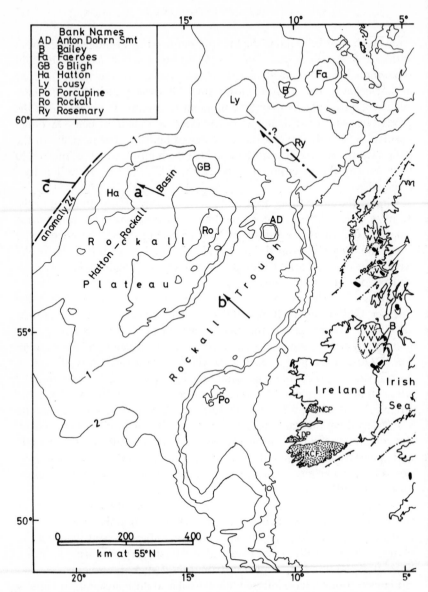

FIG. 2. Features in the North Sea and British Isles probably related to Atlantic separative phases by lateral heat flush effects. *Key:* 1. Eastern limit of erosional exhumation of deep structure; 2. Buried faults of Viking and Central Grabens; 3. Other basin outlines in North Sea; 4. Post-Cretaceous fault movements west of Great Britain; 5. Early Tertiary basalts; 6. Early Tertiary plutonic centres; 7. Plate separation directions (a and b suggested, c known); 8. Suggested transform fault

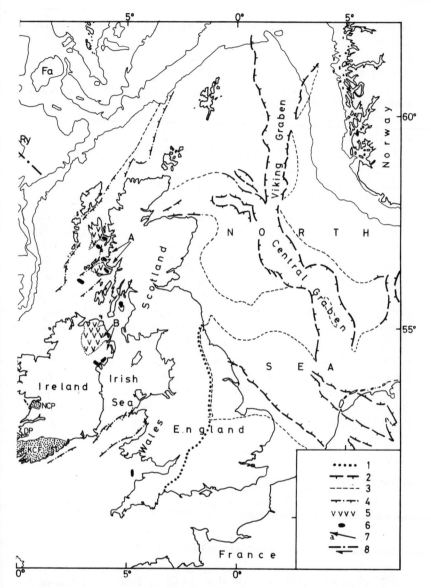

during opening of Rockall Trough. Tilted blocks in west of Ireland: KCF, Kerry–Cork Fold-belt; DP, Dingle Peninsula; NCP, North Clare Plateau. North-East strike of Tertiary dyke swarms is deflected to nearly North–South at A and B. Bathymetry (100 m, 500 m, 1 km and 2 km isobaths) after D. G. Roberts (personal communication). North Sea structure mainly from Blair.[156]

Isostatic Response to Heat-Dependent Density Changes Within Plates

General Treatment

Langseth *et al.*[58] were the first to study ocean ridge relief in terms of the heat content of plates, but they made an important error in their treatment of the connection between density increase and surface subsidence. The error has been repeated explicitly by McKenzie and Sclater,[9] Lambeck,[59] Le Pichon *et al.*[60] and Fischer,[61] and implicitly by numerous others, and must be corrected before we can go any further.

The error consists in their assumption that the lower parts of plates float passively on the material below and are not affected by subsidence of the top surface in response to density changes within the plate. Their calculations, based on the heat being lost and known coefficients of thermal expansion, then showed that the entire volume change of the rock had somehow to be concentrated in the vertical direction or, in other words, 'if contraction is not predominantly vertical the excess elevation [attributable to plate heat content on ocean ridges] would be less than 1 km and would completely fail to explain the topography' (p. 169 of Le Pichon *et al.*[60]).

The physically implausible uniaxial contraction requirement is completely avoided by the following treatment, given previously by Osmaston.[44] If a column of material contracts *isotropically*, so that its vertical length l (in this case the plate thickness) decreases by an amount x and its mean density increases by a small fractional amount from ρ to $\rho(1 + \delta)$, let the resulting isostatic subsidence of the top be h. Then, prior to contraction and subsidence, the pressure at the column's base is $l\rho$, and at the same level afterwards the pressure is $\rho(l - h)(1 + \delta)$. Equating these pressures yields

$$h = l\delta/(1 + \delta)$$

Since $x/l \ll 1$, the proportionate density change is closely equal to three times the proportionate linear contraction, i.e. $\delta = 3x/l$, which, when substituted in our relation for h, gives

$$h = 3x/(1 + 3x/l) \simeq 3x$$

The physical significance of this result is that the full isostatic compensation of density increase within a column causes the top of the column to subside by an amount nearly equal to three times the *linear* contraction of the entire column.

It is normal isostasy theory that when such primary subsidence results in a corresponding additional water or sediment load, of density ρ_L, the overall subsidence of the original top of the column is multiplied by a factor

$$M = \rho_M/(\rho_M - \rho_L)$$

where ρ_M is the uncompressed density of mantle material. Insertion of suitable values shows that M is 1·45 for subsidence under sea-water and 2·5–4·7 for sediment loading.

These amounts of surface subsidence are nearly the same as were provided by the uniaxial contraction hypothesis, so treatments of ocean floor topographic subsidence are scarcely affected. In our case, however, even without additional load, the entire column subsides and its bottom descends an amount equal to twice the linear contraction of the entire column. In addition, load-induced subsidence at the top adds directly to the movement at the bottom. This means that isostatic response to the development of thermal density anomalies within a plate *involves major vertical shearing throughout the entire plate thickness*. In the ocean floor case, the mean shearing between adjacent columns is nearly three times that implied by the erroneous treatment. Consequently, when plates are very thick we may expect their resistance to vertical shear to be evident as incomplete isostatic response to thermally induced density changes, and free air gravity anomalies will be present. When cooling has increased densities in one part of a plate, but full appropriate subsidence has been prevented by its lateral attachment to the rest of the plate, the mass excess will appear as a positive free air anomaly. The rest of the plate will be slightly depressed in providing this lateral support, so will show a negative free air anomaly. Conversely, if a heated area rises incompletely because it has to lift an adjacent unheated area, there will be a negative anomaly over the heated area and a positive one over the unheated but lifted area.

As discussed later, even under conditions of uniform cooling or heating, causative differential changes of density can arise if adjacent areas have differing constitutions, as is commonly the case on continents. It is notable, therefore, that reviews of data for the United States[62] and India[63] showed relations between elevation and gravity just like those inferred above.

We have seen that isostatic adjustments involve the motion of columns extending right through the plate. Primary thermal (and therefore density) anomalies due to plate separation or intra-plate volcanism will, however, always fade to zero at some distance (and perhaps far) above the plate base whence the causative hot material was derived. This means that inferences regarding the depth extent of such anomalies are no measure of plate thickness. On the other hand, our treatment, based on the integrated anomalous contraction (or expansion) of the column, makes a knowledge of actual plate thickness largely unnecessary.

An important corollary to our demonstration that ocean floor epeirogeny involves density increase by physical contraction that is isotropic, and not uniaxial as had been thought necessary, is that the interpretation of oceanic fracture zones as cooling cracks by Turcotte[64] and Collette[65] becomes an eminently tenable proposition, though the previous difficulty was apparently not realised by them. It also makes tenable the suggestion, particularly relevant to oceanic parts of plates, that intra-plate volcanism is caused by horizontal thermal contraction within plates. The Cameroun–Annabón volcanic line might be one good example of this; the progressive volcanism of Pacific island chains may be another.

Oceanic Gravity Anomalies and Shelf Behaviour
Irrespective of the thickness of plates, this treatment of isostatic responses to thermally-induced density changes offers, in principle, a way of accounting in detail for the development of free air gravity anomalies in areas in which overthrusting is obviously not involved. The thicker plates are, however, the greater their flexural and shear strengths are likely to be, and the greater the potential for explaining large-scale or large-amplitude anomalies. A striking feature of ship-borne gravity data and of the satellite-derived global gravity pattern (after removal of the longest wavelength components thought to be related to effects at the core–mantle boundary) is that spreading ocean ridges typically exhibit positive free air anomalies and island-free deep ocean basins negative ones, with a usual difference of 25–35 mGal. These have been attributed to a sub-plate convective flow pattern, on the grounds that a 50–80 km thick lithosphere could not support the implied stresses.[35,59,66–70] Our thick-plate model, however, not only could enable the stresses to be supported, but offers a within-plate explanation of the anomalies themselves.

Near the crests of ocean ridges there is rapid cooling and subsidence, but far down the flanks the subsidence is much slower (unless sedimentary loading is going on). If the plate between the two is structurally integral it will have to undergo continuous flexural or (if the plate is thick) shear deformation to accommodate the difference in subsidence rate. The stresses required for this deformation will result in the ridge crest being anisostatically too high and the flank basin too deep, and the observed free air anomalies are here interpreted as mainly due to this. Perhaps the well-known seismic anisotropy of the sub-oceanic mantle, which develops within the first few million years and extends from close beneath the crust to

125 km depth, and perhaps to as much as 380 km depth,[33] is also attributable to this shear deformation. The fact that between the fast-spreading East Pacific Rise and the Tonga–Kermadec Trench the gravity minimum shown by Anderson et al.[70] is only 2500 km from the ridge crest (and less than half way to the trench), may be a measure of the limited stiffness and/or relaxation time of plates, and of this one in particular. Slower spreading may produce stiffer plates because, at a given distance from the crest, they have had longer to cool.

The relation between this and previous (sub-plate convection) interpretations of the gravity data requires comment. Lambeck[59] and Sclater et al.[35] showed that part of the ridge anomaly might be attributed to any low densities beyond 75 km depth which were contributing to the compensation of the ridge. In the case of thick plates, the main thermal anomaly (relative to plate beneath flank basins) must lie at less than 75 km depth, so the contribution to the ridge gravity anomaly is probably minor. Neither this type of cause, however, nor the convection hypothesis explains the presence of apparently complementary negative anomalies confined to the flank basins, whereas our lateral support interpretation can. On the latter basis a 30 mGal anomaly difference implies that the crest-to-basin depth difference is anisostatically enhanced by about 300 m. In their detailed study of North Atlantic data, Sclater et al.[35] concluded that crest-to-basin differences were up to 1400 m too large, but that was by comparison with a cooling-slab model which allowed for neither the varying crustal thicknesses and subjacent seismically anomalous mantle near the ridge crest[71] nor the effects of a shallow top to the LV zone[32,33] during the first 20 my. Further refinement of the cooling-slab model is clearly necessary before the real excess topography can be discussed.

A direct result of our interpretation is that there is an epeirogenic connection between ridge crest and points on the same plate, perhaps to at least 2500 km from the crest. This means that, subject to this distance limitation, ocean-edge conditions (sedimentary loading, heat flush or the ensuing cooling from intra-plate volcanism, subduction) must affect ridge crests and, more importantly for the petroleum geologist, that continental margins of so-called 'passive' types are strongly coupled to any events at ridge crests (e.g. changes in spreading rate) which try to change crestal height. (Slower spreading tends to produce a higher crest.[70]) The correlations found by Rona[72] between sedimentation episodes on opposite shelves of the central North Atlantic and between these and changes in mid-Atlantic spreading rate seem to support this conclusion. The eustasy explanation put forward by Rona is also possible but awaits a proper assessment of the short-term eustatic capabilities of spreading rate changes.

It is now thoroughly established the world over that a major phase of subsidence is initiated in most parts of the subsequent continental shelf either at the presumed moment of continental plate separation or shortly thereafter. Dietz's hypothesis[73,74] of a mechanical coupling between ocean floor subsidence (which at that time he attributed entirely to sedimentary loading) and that induced in the shelf has been amply justified. In the context of our model, a physical reason for the sharp onset of the down-dragging effect is provided by the rapid cooling and density increase of the first few hundred kilometres of laterally accreted new plate. As was pointed out by Sleep,[75] the fact that the mechanical discontinuity, which is seen to have permitted uplift of the continental crust further from the new continental margin, occurred at a highly variable distance from it, poses a major difficulty for interpretations (e.g. Bott[76-78]) of shelf subsidence based on the principle of crustal attenuation, no matter whether this is attributed to plate separative tension or to lower-crust flowage under the influence of stress distributions arising from the continent–ocean change in crustal thickness. The observed sharply bounded and long-lived differential subsidence within shelves, moreover, seems inconsistent with the smudging of upper crust relations with the lower crust (and beyond) which would result from any substantial horizontal flowage there.

MECHANISM OF EPEIROGENIC DIFFERENTIATION

The term *epeirogenic differentiation* is proposed here to refer to the epeirogenic renewal or apparent initiation of the complex block-and-basin structural pattern which is such a familiar exposed and subsurface feature of the continental crust, especially those areas that have experienced Phanerozoic diastrophism of some sort.

There are strong indications that epeirogenic differentiation is at least partly of thermal origin. In the North Sea, for example, a block and basin structure which had been evident at least as early as Carboniferous time experienced two major Jurassic phases of rejuvenation and exhumation, with large ensuing differential subsidence of the basins, and has now subsided once more beneath a growing Tertiary cover.[79-83] If, as has been widely inferred,[49,51] these events are correlated with those of the Atlantic, the post-Kimmeridgian initiation of the second rejuvenative phase could well mark the start of plate separation to form Rockall Trough, although somewhat earlier than suggested by Roberts.[49,84,85] Hence the first rejuvenative phase, of end-Lias initiation, presumably marks a precursory separative movement, of which the Hatton–Rockall Basin (Fig. 2) seems a likely product. Both phases were significant thermal events for oil and gas genesis in the North Sea[51] and the first, at least, involved explicit volcanism.[86]

As suggested earlier, both these correlations and the subsequent early Tertiary one could arise from structural failure of the plate, due to uplift of the relevant separation margin, a process that would draw hot material from the plate base into any cracks so formed. That the Atlantic shelf margins were indeed uplifted is attested both by their erosion[49,84,85,87,88] and by the eastward spreads of coarse sediments.[49] In the early Tertiary event, the north-westerly strike of the dyke swarms suggests that the north–south structural failure of the plate triggered the relief of horizontal thermal contraction stresses built up at oceanic fracture zones in the floor of Rockall Trough.

Of particular interest is that individual structural elements around the south and west sides of Ireland (Kerry–Cork Hercynian fold-belt, Dingle Peninsula, North Clare Plateau—Fig. 2) show uplift which has produced strong eastward topographical tilts (up to 1:150), suggesting that they have responded to a lateral heat flush gradient from beneath the Atlantic floor whereas the intervening Carboniferous basin areas have not. Post-Palaeozoic magmatism here is confined to a mere handful of Tertiary dykes. Clearly, here as elsewhere, heat flush by itself is much too blunt an instrument to account for epeirogenic differentiation.

There are two ways of approaching this general problem. One is to contend that the differing behaviour of these often sharply delineated portions of the crust has some random small-scale cause at least as strong as the evident regional ones. The other, which we shall pursue, is to infer that the differences in behaviour signify real contrasts in the deep constitutions and histories of these small portions of the plate (in this case the Eurasian plate) which, if correctly interpreted, could lead to a greatly refined understanding of the evolution and structure of the region.

When lateral heat flush spreads into a region there are four ways in which the heat may cause a reduction of density. These are:

(1)   thermal expansion,
(2)   solid-to-liquid phase change,
(3)   solid-state phase changes from one mineral facies to another, and

(4)   reactions involving decomposition of hydrous minerals and interstitial retention of the resulting fluid.

The contributions from (1) and (2) will cause general upwarping, but are likely to exhibit sharp place-to-place variations only in the immediate vicinity of active magmatism. It was shown by Osmaston,[36] however, that the processes of types (3) and (4) likely to occur in an uppermost mantle context are from 10 to 200 times more efficient than thermal expansion for converting heat into epeirogenic movement. In the lower half of the continental crust the relationships are likely to be broadly similar, but here there is a large increase in the proportion of the material that is capable of taking part in these processes. Setting questions of isostasy aside for a moment, consider the effect of a heat-flush-induced increase of vertical temperature gradient in the region of the base of two crustal units, one of which (A) projects 10 km further into the mantle than the other (B). If conditions are right for this to cause a type (3) or (4) process both in the lower 10 km of A and in the corresponding mantle beneath B, their differing compositions will have the effect that, in raising the temperature by a given amount, much more heat for reaction will be absorbed into the base of A than into the mantle below B. Consequently, A will, if it is free to do so, experience major isostatic uplift relative to B. If isostatic adjustment is not possible, the top of A will still rise, relative to B, but by only one-third as much, i.e. by its vertical expansion alone.

The same argument, can, of course, be applied to two crustal units of similar thickness but markedly differing lower crustal constitution, but in plausible cases the epeirogenic contrast would probably be smaller. A third possible source of increased thermal epeirogenic sensitivity, but one which probably has rather blurred boundaries, is the impregnation of the mantle above subduction zones with H_2O and perhaps basaltic or andesitic material, a condition that may be retained indefinitely thereafter. Events which permit the reassertion of isostasy can also produce epeirogenic differentiation, but this does not explain the initial disequilibrium.

We conclude that differences in crustal thickness are the single most likely cause of epeirogenic differentiation in the presence of lateral heat flush but that other constitutional differences (e.g. exhaustion of water content in high grade terrains) in either the lower crust or the upper mantle may contribute substantially in specific situations. Conversely, faced in the field with sharply defined epeirogenic differentiation, the first interpretation that should be considered is one which provides for sharply differing crustal thickness. This is the subject of the next section.

GENESIS AND EVOLUTION OF CHASMIC BASIN COMPLEXES

'Simple' Chasmic Basins

The term *chasmic fault* was proposed by Osmaston[36] 'to define any major age discontinuity extending through the lithosphere' as a consequence of plate separation. It was considered that 'each such "fault" might be single or a complex fault zone, depending on the tectonic details of early separation'. Thus a chasmic fault differs from other kinds of fault in that the generative motion is separative, not shear, and in that the fault 'plane' is, initially at least, a welded igneous contact. In this sense, therefore, most of the Atlantic Ocean basin lies between chasmic faults and may be regarded as a chasmic basin of very large size. For our purpose a 'simple' chasmic basin involves only one overall separative movement.

It was shown[36] that if separation is more limited in amount the known 3–4 km cooling subsidence of ocean lithosphere, combined with the isostatic response associated with sedimentary filling of the basin to sea level, would, on reasonable assumptions, eventually result in a basin crust with a thickness of about 26 km. The much increased cooling time and greater total subsidence implied by our conclusion that plates are very thick means that total subsidence times may, in fact, reach 500 my or more and result in crustal thicknesses of 30–35 km. This makes the concept[36] that chasmic basins can become part of the continental crust, without the need to invoke lateral squashing of any kind, an even more attractive proposition than before.

The Tectonics of Separation

Notice at once that, if the not infrequently small scale of the blocks and basins involved in epeirogenic differentiation has an explanation (as the author believes) in the different crustal thicknesses and compositions resulting from complex sequences of plate separation, it is unlikely that the basins contain any substantial amount of foundered crust that is otherwise the same as the blocks. The apparently direct conflict here with the evidence from continental rift valleys is dealt with later. Just as for intra-plate volcanism, thick plates make magma genesis (but not necessarily magma extrusion) an inevitable consequence of even the smallest amount of plate separation. However, the rate of separation makes an immense difference to the structural consequences.[44] At the relatively high separation rates typical of relative plate motions (say 6 cm/year) calculations show that although, at the very beginning, lateral heat loss will severely reduce the ease with which the magma reaches the surface, this will very quickly be overcome and, by the time separation has attained about 2 km, the axial temperature will be close to its steady-state separation value. Consequently, long before the separation has got this far, the crust on either side will be substantially supported against lateral collapse by accreted and upwelling magma of comparable density. The accretionary ridge may even build considerably above the adjacent continental crust before lateral heat flush has penetrated far enough to cause uplift; indeed, Sheridan's sections of the Atlantic shelf of North America[90] suggest just this, in that the outermost basement of the continental shelf is now bordered by a ridge of oceanic basement which rises far above it. In view of this, the author recently proposed[91] that these are, in fact, the circumstances in which ophiolites are initially emplaced (by sliding) onto continental margins, a matter whose tectonic importance is discussed later. Finally, notice that the East Pacific Rise, west of South America, has a high separation rate but no axial rift.[44]

In conclusion, provided the new margins were in isostatic equilibrium beforehand, and provided that the rate is high, continental splitting is likely to produce sharply defined chasmic faults at depth with only slight (1–5 km?) loss of crustal map outline near the surface (beneath any transgressive cover). This important result is supported by the remarkably close-fitting reconstructions which have been found to be required by the data in actual cases (author's unpublished work). Previous treatments (including one by the author[44]) of this problem have always started from the probably mistaken premise that continental rift valleys, with their extremely slow rates of opening and consequently different thermostructural regime,[44] represent the first stage of major separation. Moreover, the structural stiffness, which, as we have seen earlier, is evidently possessed by tectonic plates, makes implausible the popular notion that tensional necking of the plate thickness is a normal prelude to separation. This conclusion in no way denies, however, that normal rotational faulting of the *upper* crust can and does frequently occur when it is deprived of lateral support.

Basin Subsidence
The size of the basin produced fundamentally affects the subsidence process in several ways. In very wide basins (e.g. the Atlantic) the down-drag of the enormous cooling floor dominates processes at the continental margin and produces the deceptive similarities to half-grabens. In basins which are sufficiently small when opening ceases, cooling of the sub-floor plate material will be accelerated by lateral heat loss to the adjacent cooler

lithosphere. Sedimentary loading with consequent thermal blanketing will become major factors. Furthermore, the smaller the basin, the more will isostatic subsidence of the floor tend to be restrained or inhibited by its welded attachment to the adjacent lithosphere. Consequently, in the absence of external tectonic disturbance, relative subsidence of the basin may cease, enabling its outline to disappear beneath a sedimentary blanket, only to reappear suddenly when isostatic response to the continuing density increase at depth becomes possible. This process probably contributes substantially to the differential movement which accompanies epeirogenic rejuvenation. Both here and in the case of lateral heat flush the older, thicker crust rises and the younger, thinner crust subsides/rises less. Down-faulting will not necessarily follow chasmic fault lines, but will depend on the overall stress distribution and, because of the horizontal contraction of the basin sub-floor material, will tend to dip steeply basinward. Clearly there is much here that accords with experience.

Basement Structure of the Basin

There seems every reason to expect that oceanic-type fracture zones, in their capacity as cooling cracks, will be generated, especially when the separation is quite limited (say 100 km), because of the influence of irregularities in the separated outlines and because of more vigorous lateral cooling effects. Such fracture zones will give the basin a characteristic basement grain which will provide permanent loci for differential movements (and mineralisation?) within the basin and whose orientation is a direct indication of the separative direction involved in the basin's genesis. The prominent south-easterly basement grain seen to control gas field and other structures in the basins of the southern North Sea,[51] seen again in the English north-east Midlands[92] and in eastern Germany, runs parallel to the well-known Tornquist line, marking the edge of the Russo–Baltic shield, and could have such an origin. It clearly influenced strongly the late Hercynian wrench directions.

The most likely magnetic and gravity lineation corresponds to the fracture zone grain. Non-generation of any sea-floor spreading type anomalies in the presence of active sedimentation is indicated by the fading-out of such anomalies under even quite thin sedimentary cover in the Red Sea and the Gulf of California, and appears to be due to inhibiting the usual rapid chilling by sea-water, although the exact mechanism is not clear.[36,93]

Island Inliers

In the past these have been called micro-continents, but the author prefers the term *island inliers* in view both of the very small size of some of them and of the structural form in which they appear when surrounding sedimentation has consolidated them into the continental crust. The traditional view of inliers, that they are the best available evidence of what lies buried in the neighbourhood, must be abandoned for structural islands in continental areas if we wish to account satisfactorily for basin subsidence, just as their distinctive nature has been accepted within the ocean basins. The detachment of island inliers requires that the axis of separation lies first between the inlier and plate *A* (say), then jumps to start separating the inlier from plate *B*. Thus the inlier is always attached to one plate or the other and the kinematics of its displacement depend precisely upon the relative motion of the two major plates. There is no evidence that independent motivation is available to any plate that is not at least 1000 km across, unless one of its margins is an active subduction zone. This places valuable constraints on plate tectonic reconstructions. Because island inliers are at all times part of a major plate there is no question of structural instability, even for very small inliers. The East Canary Islands block[94] is an example of an island inlier with a morphologically well-defined relation to its adjacent continent. The intervening salt basin is of Late Triassic age[95] and ophiolite is possibly present on Fuerteventura.[96]

Effects of Compression
If folding, thrusting, or the emplacement of allochthons occurs while thermal subsidence of the basin is still in progress, the continuance of subsidence thereafter will help erosion to level any subaerial edifice produced, and hasten the return to a basin regime. This is important, as folded and eroded basements are by no means uncommon in deep basins. For example, Carey's proposal[97] that the West Siberian lowlands are a filled rhombochasm (= chasmic basin) was rejected by van Andel and Hospers[98] on the grounds that the deeply buried (up to 8 km) pre-Callovian basement is folded,[99] but it is unlikely that the subsidence of this huge area can be explained in any other way.

The imposition of a horizontal stress field upon a complex of chasmic basins will promote failure and down-faulting near basin margins where any substantial vertical shear stress had already built up. A probable example is the late Hercynian (basal Permian and end Scythian ('Hardegsen')) down-faulting of the North Sea and western Britain.[51]

Polyphase Separation
Nearly all 'passive' continental margins that have been investigated

sufficiently show evidence that a basin or basin complex existed there long before the oceanic phase of separation. Where, on reconstruction, the basins are seen to have been of limited width between stable platforms, cooling subsidence of their floors could well have been much hindered by lateral attachment to the lithosphere of the platforms. Relief of this disequilibrium at the start of the oceanic separative phase will involve both a down-faulting of the basin floors (before the onset of ocean floor down-drag) and a matching load-relief uplift of the adjacent platforms, possibly long before lateral heat flush from the ocean floor has had time to reach the latter. These considerations show the great importance of the pre-existing basin in interpreting the overall behaviour of the margin. In some cases not two, but three or more, separative phases must be considered. Along the Niger–Cape Town margin, for instance, data taken from Emery *et al.*[100,101] and references therein, and from Rigassi,[102] suggest that three phases are relevant, namely,

(1) Triassic or earlier, possibly of rift valley character,
(2) late Jurassic rapid but limited separation,
(3) early Aptian oceanic phase causing thick evaporites on the floors of (1) and (2) in a basin initially barred to the south by *en echelon* island inliers (Walvis Ridge, Rio Grande Rise).

In such a case, full structural interpretation requires the use of reconstructions with the conjugate margin, and should be helped by recognising

(a) that the (1)/(2) and (2)/(3) boundaries in the basement are likely to be sharply defined (albeit superficially indistinct) chasmic faults, and
(b) that detachment of island inliers may have occurred on either occasion.

Similar complexities will arise in chasmic basins of polyphase origin which became incorporated into the continents because none of the phases was of oceanic magnitude. Unless there is a large proportionate age gap between phases, the chasmic faults which separate parts of the floor created during different phases of separation may now be very poorly marked in comparison with those that define the basin as a whole. Within the limits set by the overall geometry of the interacting plate margins concerned, successive phases may differ markedly in direction, with the result that a single-phase overall interpretation of the basin would yield little of value for the interpretation of basin floor or margin-related structures.

Finally, continental grabens in which volcanics are scarce or absent (Rhine–Rhone system,[103] Baikal system[104]) may, perhaps, be properly regarded as two-phase basins. Interpretations of these, based on an ancient chasmic basin phase, followed by a much later slight separation which admitted magma at depth and induced major epeirogenic differentiation and some flank volcanism (by mechanism (b) of intra-plate volcanism proposed on page 10), would help to explain why the slow-single-phase (?) Gregory Rift in Kenya[105] exhibits such a contrasting structural and magmatic style.

FEATURES RELATED TO APPROXIMATIVE PLATE MOTIONS

Plate tectonic analysis of basin systems interspersed in orogenic belts requires that the effects of approximation be carefully distinguished from those of separation. We now consider briefly how the concepts developed in preceding sections affect the consequences and interpretation of approximative motion. Apart from those cited below, further relevant papers will be found in Burk and Drake.[106]

Subduction-related Processes (see also the note added on page 51)
The steep dips of Benioff zones imply severe deformation of the previously horizontal plate as subduction proceeds. Most authors have assumed a bending mode of deformation but Lliboutry[107] pointed out that this involves a second deformation, to unbend the plate after acquiring the necessary dip. He proposed instead, as being consistent with seismic first motions in the down-turn zone, that deformation is by vertical shearing, a mode which becomes physically more likely the thicker plates are, owing to its lower energy requirement, and seems certain in our thick-plate context. However, the topography does indicate slight initial bending (3°) during the approach to some trenches and it is perhaps this which causes the observed erasure of the oceanic magnetic anomaly pattern. The main down-turn zone coincides with the strong free-air gravity minimum, which lies at up to 150 km to landward of the trench axis.[108] At the down-turn line, successive slices of plate will be either pushed down on encountering the overlying thrust plane, or dragged down when the slice in front encounters it, producing an effect on the oceanic plate closely analogous to that seen at the top of a descending escalator.

This model has interesting properties. The steps thus produced on the top surface of the down-going plate must constitute a continually renewed means of eroding the overlying (landward) side plate unless these 'teeth' are equally continually clogged with sediments. If the supply of these (including stratified material on the oceanic plate) is more than enough to achieve this, a wedge of terrigenous sediment will spread across the down-turn zone onto the relatively undeformed surface of the approaching oceanic plate. This sediment will then probably be formed into a folded and imbricated structure, attached to the overriding landward plate by repeated 'under-tucking' of the distal portion of the wedge, as in the observationally well-supported model of Seely et al.[109] and Mitchell and Reading,[110] to form eventually a blueschist-metamorphosed fore-arc outer ridge. When this happens, the 'appetite' generated by step-formation at the down-turn zone might be satisfied by down-faulting or subsidence of the imbricated structure above the zone, to produce a terrace or inter-arc flysch basin. It is not clear whether the build-up could become massive enough to promote earlier down-turn of the oceanic plate, to produce the kind of systematic oceanward migration of belts seen in south-western Japan, or whether this requires that subduction be stopped and restarted. Where sediment supplies are small the topographic trench will lie at the down-turn zone, so it will tend to be deeper and have a steeper (and stepped?) oceanward side. Any tectonic erosion of the hanging wall of the thrust plane by the down-going steps may move the volcanic front correspondingly inland, as in Peru and Chile.[111-113] Karig[114] doubted that tectonic erosion occurs anywhere.

Notice that this model of the down-turn process requires that, when the oceanic plate carries only a thin layer of previously deposited sediment, a substantial volume of predominantly landward-plate derived material, consisting of sediments or of subterraneously eroded pre-existing crust, should be continuously carried into the Benioff zone by the oceanic plate. Long-term recycling of these materials will occur when they are melted (together with oceanic crust) to form magmas erupted from the Benioff zone, possibly giving rise to continental-type trace element patterns. Consequently, the finding of high $^{87}Sr/^{86}Sr$ ratios in the calc–alkaline volcanics of Java, Japan and the central Andes (especially the latter)[26,115] gives support to the model.

The stresses required for the down-turn deformation of the oceanic plate make it inevitable that a substantial part of the large free-air gravity anomaly couple (often of 200–400 mGal peak to peak) associated with active subduction systems is due to the consequent mutual reaction between the two plates, causing the landward plate to be supported substantially above isostatic equilibrium. McKenzie[66] considered this to be a possibility even if plates were only 100 km thick. Gravity-controlled interpretations of crustal thickness contrasts across trench systems need to recognise this if they are to give a true picture. This also means that, for purposes of palaeo-tectonic analysis, the onset of active subduction should be indicated geologically by substantial (perhaps 500 m) uplift of the landward plate. This mechanically applied uplift force at the plate margin, supplemented by thermally-induced buoyancy at the volcanic belt, must produce a large concave-upward bending moment upon the rest of the plate lying to the rear of the volcanic belt. This is likely to be an even more powerful cause of plate structural failure than that which we have already inferred to be effective in the case of buoyancy induced at separate plate margins, and could well be the mechanism of plate rupture which permits the opening of rear-arc basins. Positive free air gravity (15 mGal average) over rear-arc basins[116,117] implies that such basins, integrated once more into the plate structure, share in its uplift.

Rear-arc basins seem to have opening kinematics very like other chasmic basins. Notice that subduction is clearly irreversible, whereas the (endwise?) subduction of actively spreading ridges could very easily produce temporarily a demand elsewhere for compensatory plate production in lieu of consumption. For example, Japan experienced a lull and reorganisation of subduction during the late Oligocene and early Miocene, which is probably when the Sea of Japan was opened.[118,119] It appears that arguments (e.g. Karig[120]) that rear-arc basins have opened *while subduction was active* (except, perhaps, in a mutually perpendicular direction) need to be scrutinised with extreme care.

Collision Processes

In the preceding section we saw that, as the volume of sedimentary material presented for subduction at a trench increases, the process passes from tectonic erosion of the overriding plate to tectonic accretion. When the thickness of sial presented for subduction rises further to continental values, as when a passive continental margin arrives at the trench, collision is said to occur. Depending on the motivating drive of the plates concerned, subduction may be halted, or it may be only slowed, bringing into play major deformational forces at crustal levels. The upper half of the crust on the downgoing plate will tend to get sheared off and pile up as folds or imbrics (e.g. Zagros[121,122] and Himalaya[123]), constituting a major development of the previous accretionary process. The overriding plate margin, on the other hand, will tend to experience not only compression but vigorous uplift and tilting of its surface away from the collision zone. If there is a thick overlying sedimentary section (which is unlikely if prolonged subduction has preceded collision) this tilting may initiate the decollement of Alpine-type nappes. This reasoning differs from that of Dewey and Bird,[122] who believed that ocean-floor subduction, without collision, could cause nappes.

Interpretation of the behaviour of colliding margins frequently depends closely upon one's interpretation of the previous (often separative) history of the margins concerned. Ophiolites, in their capacity as slices of crust generated during oceanic-type plate separation are of well-recognised importance in this respect. With a few notable exceptions[124-126] most previous authors have invoked some perversion of normal subduction for the emplacement of ophiolites. However, the author previously,[91] and again in this chapter, has developed Church and Stevens' suggestion[126] of emplacement during early separation, by distinguishing initial emplacement on to the margin from the usually much later subduction- or collision-related final displacement. This appears to overcome the six main difficulties of subduction-related initial emplacement, namely:

(1) A high enough source suitably placed for emplacement by sliding (major examples are little deformed and rest on an undeformed autochthon);
(2) The common occurrence of thermal aureoles;[127-129]
(3) The short life span before emplacement;[130,131]
(4) Synchronised emplacement at many points along the relevant plate margin;[130,132]
(5) Occurrences of identical age on opposite sides of a basin or suture (e.g. Oman–Makran); and
(6) Emplacement typically very early in the particular orogenic cycle concerned.[129,133]

For purposes of tectonic analysis this provides some separative events, at least, with a well documented time marker, enabling advances to be made in the study of alternating separation and approximation in orogenic belts.

The rather general belief that chasmic basins are by their nature fated eventually to undergo major crustal shortening or total subduction appears erroneous. Unpublished tectonic analyses by the author show that subduction can be halted when a substantial island inlier or group of inliers, detached during earlier separative movements, reaches the trench. The chasmic basins behind it are thus protected and enabled to mature into continental basin crust. The occurrence of major crustal shortening appears to depend on either the possibility of major décollement and imbrication (south-west Zagros, Himalaya) or the thermal softening-up which occurs in the magmatic belt above Benioff zones. It is very notable, however, that wherever we see subduction of oceanic plates occurring, deformation of the magmatic belt is strictly germanotype, horizontal deformation being minor or absent.[134,135] This implies that major deformation of the high temperature metamorphic belt of an orogen must be restricted to the collision stage, a matter of some importance in the analysis of orogenic belts.

For example, the occurrence of probably late Cambrian folding and metamorphism (Grampian/Fleur de Lys/Caldwell) in the Caledonian–Appalachian belt makes it unlikely that the same Cambrian proto-Atlantic ocean continued to close thereafter until the early Devonian, as proposed by Dewey[130,136] and adopted by numerous others. This, in turn, means that post-Cambrian separative phases must have occurred to provide the crust which was consumed during the explicitly subductive phases (granites, calc–alkaline volcanism) which occurred in Llandeilian–early Caradocian (Taconian phase) (for example references 137 and 138) and early Devonian[139] times. The early Ordovician initial emplacement of ophiolites all along the belt[129,140,141] evidently marks the first of these separations and explains why the corresponding sediments escaped metamorphism at this time. Wherever the separations involved fragmentation some of the chasmic basins produced may have survived. It appears probable, for example, that the Carboniferous block and coal basin structure of the Scottish Midland Valley is entirely the product of these earlier motions.

Turning now to the Alpide belt, there is some evidence[123,132,142–144] that the original northern margin of Gondwanaland now passes between Iran and the Greater Caucasus, and thence eastward via the Hindu Kush to a line (Kunlun Mountains) along the northern border of Tibet, a position attained by southward subduction beneath this margin. Based on data and citations given by Stöcklin,[142] Stoneley[132] and Le Fort,[123] the author proposes the following interpretation.

In late Cretaceous (Maastrichtian) time much of Turkey, Iran (north of the Zagros crush zone), Afghanistan and Tibet (north of the Indus suture) became detached northwards from Arabia and India, causing emplacement of the 'South Tethyan' ophiolites (Taurus, Troodos, Hatay, Neyriz, Indus suture) along the new margins. (The Zagros rift had a separative history going back at least to the late Triassic, which previous workers have regarded as the time of the main separation.) The slightly older (Coniacian) Semail ophiolite sheet of Oman rests on the immense Hawasina pile of nappes.[125,131,145] These appear to be the cover rocks from the extensively exposed Infracambrian basement, and neighbouring areas, in the western part of the Lut-Nain block of east-central Iran, whose faulted margins are fringed with Campanian-Maastrichtian ophiolitic melanges (Stöcklin[142]). The Semail ophiolites must have been generated north of the Lut-Nain block (Dasht-e-Kavir chasmic basin) and slid southward onto it, inducing décollement of the cover rocks, probably on the Infracambrian salt (Hormuz correlative) which is widely present beneath cover rocks elsewhere in the region, as the block tilted southward due to lateral heat flush. The assembled allochthon then reached Oman just before the Lut-Nain block split from it in the Maastrichtian. This approach to the solution of the Hawasina source problem would be impossible without the new interpretation of the ophiolite emplacement process. The associated flysch sedimentation is discussed later.

The northward-moving blocks (Turkey, Iran, Afghanistan, Tibet) may have collided (not very forcefully) with Eurasia in the early Palaeocene (Elburz folding). The final approximative movements by India (throughout the Eocene[11,123]) and Arabia (rather later) involved north-eastward subduction beneath Tibet and Iran, producing first folding (initial collision) in the Himalayas in the Middle or Late Oligocene,[123,146] but not until the Late Miocene in the Zagros south-west of the Crush Zone suture—a well-documented fact[121,147−151] which has two important implications.

One is that the tectonic quiescence of the south-west Zagros part of the Arabian plate in the Lower Miocene was such as to permit deposition of the remarkably widespread and uniform evaporitic Gachsaran (ex-Lower Fars) Cap Rock formation, beneath which more than 10% of the world's known petroleum resources[152] have remained sealed despite the intense fracturing of underlying formations caused by the subsequent folding. The other implication is tectonic and concerns the opening of the Red Sea. There is much evidence that a Red Sea separative phase occurred at some time during the end-Eocene–mid-Miocene interval, but if Arabia and Iran had already collided in the Late Cretaceous (for example references 142 and 151) this would have raised an important question as to the sufficient motivation of the comparatively small Arabian plate. However, our inference that an oceanic subduction zone existed at this time along the (then) south-west margin of Iran means that Red Sea opening was relatively unimpeded. Motivation of the Arabian plate probably came from the north-eastward rubbing motion of the Indian plate along the ancestral Owen–Murray fracture zone, off the south-east coast of Arabia. Lateral heat flush from the widening Arabian Sea probably caused the Oligocene and Lower Miocene transition from shale to the partly neritic Asmari Limestone and then to the Gachsaran evaporites. During the Pliocene–present Red Sea opening phase,[153] Arabia and India appear to have behaved as one plate.

The main source of Iranian and Tibetan uplift appears to be the driving of the Arabian and Indian shields down the Zagros and Transhimalaya subduction zones, for the uplift is sharply reduced directly north of the Gulf of Oman, where closure of the 'Southern Tethys' is not yet complete.[132,154] To produce the uplift it may not be necessary, in the context of thick plates, for the subducted shield to melt and percolate upward into the overlying plate. It would be quite sufficient that the low-density material should be present within the combined plate thickness. At the slow subduction rate (2 cm/y?) likely during this collision, fusion of the subducted crust may have been quite limited, but presumably enough to prevent its conversion to a dense eclogite facies.

Tibet has been uplifted 4–5 km by India's continuing northward drive during Miocene–Pleistocene time.[123,146] Some 2500 km to the north of the Himalayas, the extensive Baikal Rift System and related flank volcanism and epeirogenic rejuvenation,[104,155] which was initiated suddenly in the Oligocene and intensified in the Pliocene, seems very likely to have been caused by structural failure (in bending) of the then-integral Tibetan–Asian plate when its southern edge was raised. Both the Himalayan belt and the Baikal Rift system are seismically active, suggesting present-day continuance of the northward drive by the Indian plate. The distance involved is at least twice the leverage distances inferred on p. 13 for plate failure due to lateral-heat-flush-induced marginal uplift, which suggests that the compressive stress provided by India has helped to prevent tensional failure of the lower part of the Asian plate. This interpretation means that the spectacular uplift of Tibet, and of the Tsaidam and Tarim basins to the north, depends on the continuance of Indian thrusting motion. When this ceases, not only will the mechanically derived uplift force disappear but the crust on the subducted Indian plate will be enabled to convert to eclogite facies, leaving the slowly-dying thermal anomalies and any local crustal thickening as the only remaining sources of uplift. In that event, the basins to the north may quite rapidly resume marine sedimentation, as happened in the southern North Sea, following the similarly-caused uplift (and local compression) at the time of the powerful Maastrichtian compressive phase in the Alps.[81,82]

IMPLICATIONS FOR SEDIMENTATION AND HYDROCARBON LOCATION

It is intended here to underline some of the ways in which the aspects of plate tectonics that we have been discussing can be related to the interpretation of sedimentation, to the incidence of heat flow in basins, and to the investigation and interpretation of subsurface structures, having particular regard to their bearing upon the occurrence and the finding of hydrocarbon accumulations.

Plate tectonics can be applied in hydrocarbon exploration, firstly, in identifying and building up a picture of the plate tectonic regime and, secondly, in understanding the basement structure and its behaviour under the influence of that regime. This combination, it appears, largely controls both the actual sediment distribution and the structural development. The study of sedimentation in relation to tectonic regime is thus of enormous importance, not just in respect of those deposits which might provide hydrocarbon sources, reservoirs or seals, but in the identification of the sequence of plate motions and tectonic regimes that have previously generated the basement structure of the region concerned.

From the now considerable literature attempting to relate sedimentation to plate tectonics, or to features having a recognised plate tectonic significance, a few general treatments,[110,157-160] and especially those of Dickinson[161] and Dott,[162] are noteworthy. However, whilst all such authors have recognised that plate separation is a prime way of generating sinks for sediments, relatively scant attention has been given to the now-evident orogenic/epeirogenic capabilities of plate separation in providing major sources of sediments and in controlling their detailed character and distribution. The term *separative orogeny*[36] is surely appropriate to these capabilities, in view of the many mountainous uplifts which today border separative continental margins, but whose present elevations (up to 3 km or more) long post-date their structural evolution and subsequent planation.

Sediments generated by separative orogeny run a considerable risk of being confused with those from approximative orogeny in orogenic belts where, as is commonly the case, both separative and approximative plate motions have taken place. In north-western Europe, the late Jurassic eastward spread of clastics from sources near the shelf edge[89] involves no such ambiguity. Nor does the coastal-deltaic complex, which prograded eastward from the Scottish Highlands in the Palaeocene.[164] This body now provides the reservoirs for the Forties and Montrose oil fields, and was clearly caused by the uplift and still-evident eastward tilting of the

Highlands, produced by the heat flush from the Palaeocene magmatism on the west coast (Fig. 2).

The deposition of flysch (a term which Kay[165] has stressed should refer to the depositional characteristics and not to the orogenic context) onto the young floors of chasmic basins is widely accepted as certain, but it may also be caused well 'inland' from the separation margin if suitably deep troughs occur there, as our Middle East example, discussed earlier, shows. There, along the south-west side of the Zagros Crush Zone[147-151] and in Oman,[125] are the remains of a trough which received thick Senonian[147,151] cherty flysch, exotic blocks and ophiolites from the north-east. This trough was probably a limited-width chasmic basin of Jurassic or earlier age, representing the earlier rifting history of the Zagros suture, and was caused to down-fault (much as the North Sea grabens did in the Jurassic) by incipient structural failure of the Iranian–Arabian plate when separative activity in northern Iran upwarped the plate's north-eastern edge. As described earlier, this incipient failure then developed into full separation in the Maastrichtian, after which such floor of the half-graben as remained attached to the Arabian shield acted as a continental shelf until the Miocene collision. It appears that during this interval occurred much of the structural growth of basement features over which the giant oil fields of Arabia and Kuwait are located,[149,152] presumably owing to lateral heat flush effects.

At the other end of the tectonic activity scale, notice that the gently subsiding environment of a well-filled and sufficiently aged chasmic basin would provide, and apparently did in the Carboniferous, an ideal environment for the accumulation of thick coal deposits. The extensive coal-derived gas accumulations in the southern North Sea therefore favour a chasmic origin for this basin. Evaporite deposition is, of course, an obvious possibility in any restricted chasmic basin, but not in high palaeo-latitudes. Basins formed over shields, however, can also 'go evaporitic' as a result of marginal upwarping by lateral heat flush, as suggested earlier for the Gachsaran evaporite of the south-west Zagros. The widespread involvement of evaporites in oilfield situations, both as seals and as local tectonic agents, needs no emphasis here (see Halbouty et al.,[166] for example).

Heat Flow in Basins

The favourable influences of moderately elevated temperatures during petroleum maturation and migration have been discussed by Klemme[168] in relation to observed thermal gradients in basins. In the case of sufficiently young chasmic basins (e.g. the Red Sea and Japan Sea) high heat flows are due to lithospheric cooling. Similarly, high heat flows are to be expected among the active or not-too-long extinct magmatic belts of plate approximation zones. Elsewhere, intra-plate magmatic injection or extrusive volcanism must be the main source of any extra heat, and will usually be marked by strong epeirogenic differentiation with an uplift bias. Such situations are readily noted in the present, but their detection in the past (e.g. the North Sea in the Jurassic) may demand much geological investigation if there are no plate tectonic arguments to provide guidance and controls.

Subsurface Features
We turn now to the origin and behaviour of positive subsurface basement features, so many of which control the location of major oil fields. The great majority of these appear to have originated as isolated or partially-isolated island inliers or as marginal promontories within chasmic basins, and range upward in size from a few kilometres across.

The justification for this inference is as follows. The important characteristic of an island inlier in a freshly-generated chasmic basin is its much greater crustal thickness. This means that, until the basin floor has eventually attained a similar thickness by subsidence and sedimentation, the top of the inlier would always, if isostasy were to prevail, stick up above the surrounding sedimentation surface. However, just as we argued earlier that a small chasmic basin floor will be prevented by its marginal attachments from subsiding freely, so also an island inlier that is small in relation to the basin will tend to be dragged down with the surrounding basin floor. This has recently[167] been well demonstrated in the oceanic realm from Deep Sea Drilling Project data for a number of positive features (including Walvis Ridge and Rio Grande Rise) which the author believes to be island inliers probably possessing pre-oceanic chasmic basin crust. The data suggest that these inliers in fact subsided a few hundred metres less than the surrounding floor, perhaps by means of vertical shear-creep distributed across a zone of the basin basement near the inliers. This effect would be increased in smaller chasmic basins, where heavy sedimentation results in much larger depression of the chasmic basin basement, and the associated depression of the inlier produces larger buoyant forces upon it. Consequently, strata will thicken basinward from the inlier, producing near the inlier the familiar steepening-with-depth dip pattern that is characteristic of structural growth.

With increasing total burial of the inlier, structural growth may gradually stop, but this situation has large built-in vertical shear stresses at depth; any superposition of extraneous tectonic stresses may cause failure by faulting at or close to the inlier margins, enabling the inlier to 'pop up' towards isostatic equilibrium. If release faulting does not occur all round the inlier, the inlier will tilt away from the released edges. Sediments overlying the inlier will be uplifted and will be eroded if the basin was shallow. The fact that this stress release does not need to have major tectonic accompaniments in the vicinity means that subsidence of the basin may continue quietly thereafter, probably taking the inlier with it once more and bringing about burial of the island (formed above the island inlier) beneath a post-unconformity sequence of shales and/or carbonates.

This sequence of events is almost perfectly reproduced in the intimately-known structure of the Oklahoma City oil field.[169] Here it appears that a roughly triangular island inlier of Precambrian crust, not more than twelve kilometres long and six wide, forms one of a number of buried uplifts in a chasmic basin complex of probably latest Precambrian age, whose geology[170] has been interpreted as an aulacogen by Hoffman et al.[171] The Cambrian–Mississippian basin sequence accumulated at a decreasing rate, with 500 metres of Pennsylvanian structural release along only one side of the inlier, producing erosion and tilting of the above-inlier surface. Oil production is from beneath the unconformity, both on the crest and down the flanks, and from post-unconformity strata domed over the inlier by a small amount of further structural growth.

The buried Golden Lane Atoll, Mexico, an ellipse 150 km long, around which the many Golden Lane oil fields are located,[172] is probably also ultimately based on an island inlier which, in the mid-Cretaceous, defined a shoal area upon which the reef started to grow.

The probable importance of island inliers as controls upon the location of trapped hydrocarbons is thus clear. The edges of chasmic basins involve conditions similar to those at the edges of very large island inliers, in that the down-drag by the basin floor is in this case resisted by the buoyancy of a much larger slab of crust. The inherently contrasting epeirogenic properties of the basin floor and the bounding older crust imply a considerable potential for trap formation. In two-separative-stage chasmic basins, island inliers made of first-stage chasmic basin crust may provide important reservoir and trapping facilities. The foregoing account of island inlier evolution related to only one phase of stress release, but stress build-up and further faulting events may well be possible, particularly if the first phase does not occur too late in the subsidence evolution of the basin. In view of the sharpness with which chasmic basins appear to be definable if one applies the interpretive methods discussed earlier, it appears that plate tectonics can contribute, not merely to the understanding of basin edges and island inliers when found by other means, but also, by the use of rigorously precise tectonic reconstruction, can contribute much toward the location and definition of such features.

REVIEW

The plate model outlined here differs fundamentally from the one widely accepted only in respect of the plate's much greater thickness, but the consequences impinge upon surface and near-surface phenomena to a quite remarkable degree and in ways that seem to resolve a number of important previous difficulties. In particular, it offers major advances in the interpretation of basins, and of structures within them, both in orogenic settings and elsewhere.

The material of the LV zone, with a typical fluid content of perhaps 1 %, is included in the plate thickness. The sharp dependence of its thermal resistance upon its fluid content is inferred to stabilise the heat flow through it, enabling (beneath oceans) the material above it to cool almost independently for the first 80 my, as observed. Tentatively, in the light of various ocean–continent differences at depth, plate thickness ranges from 250–300 km beneath oceans to more than 400 km beneath parts of continents. The primary consequences of the greater plate thickness fall into two classes, thermal and structural.

The thermal consequences stem from the higher temperature at the plate base, which

(1)   makes intra-plate volcanism a simple consequence of cracking the plate;

(2)   greatly increases the heat brought into the plate and lost to the walls as lateral heat flush (to produce epeirogenic uplift) during volcanism or larger amounts of separation;

(3)   increases both the total cooling time for a young plate to several hundred million years and its total subsidence under sedimentation enough to give 35 km crustal thickness, making chasmic basins within continents a feasible proposition;

(4)   raises the height of embryo 'ocean' ridges during the fast splitting of continental parts of plates, so that the ridge laterally supports and may overtop the adjacent margin, enabling ophiolite slices to slide onto the margin at that time (see also Osmaston[91]).

The structural consequences stem from the much greater plate strength and flexural rigidity than previously recognised. This is evident

(a) as large-scale free-air gravity anomalies over oceans, indicating the transmission of structural support between fast- and slow-subsiding parts of plates as much as 2500 km apart,

(b) as wide upwarps (Arabia >700 km) when lateral heat flush lowers densities for a short distance inward from one edge,

(c) as structural failure of the plate (many examples) at distances of 500–1500 km from plate edges uplifted as in (b), with various rift-like and volcanic consequences, and

(d) as structural failure (Baikal) of the plate 2500 km from the underthrust uplift of Tibet.

It appears also that the subsiding floors of large chasmic basins (e.g. oceans) build up enough down-drag to pull down their margins with them, whereas the floors of small chasmic basins (including those kept filled with sediment) tend to get stuck until tectonic disturbance enables them to drop towards isostatic equilibrium, producing grabens or half-grabens. At the entry to subduction zones, thick plates favour an escalator type of down-turn deformation, rather than bending.[107] This explains evidence that the usual tectonic outbuilding gives way, in particular circumstances (e.g. the central Andes), to tectonic erosion and recycling of continental materials.

Correction of a widely overlooked error in treating the isostasy of tectonic plates, shows that non-isotropic (i.e. vertical uniaxial) thermal contraction is *not* needed to explain ocean floor subsidence or any other thermally-induced epeirogenic movements. Consequently, horizontal contraction effects are very important (fracture zones, volcanism, normal faulting).

Epeirogenic differentiation in continental and shelf areas, involving enhancement of a pre-existing block and basin structure in the presence of a thermal event, is directly attributable to sharply bounded differences in crustal thickness and (to a lesser extent) constitution. Both the implied relative thinness of the basinal crust and its subsidence-prone character are better explained as the consequence of being chasmic basin crust which is immature (i.e. sedimentary thickening is incomplete) than by crustal attenuation arguments.

Small, rapidly opened, chasmic basins

(1) will not usually exhibit oceanic-type magnetic anomalies;

(2)   will exhibit a fracture-zone-oriented basement grain which betrays
       the generative separation direction;
(3)   will subside episodically, rather than continuously, along faults
       that are commonly basinward of the genetic chasmic faults;
(4)   will often surround island inliers (the micro-continents of Dewey
       et al.[173]) upwards of a few kilometres across, but only the largest of
       these will exhibit fully independent epeirogenic behaviour.

Basin floors generated at different times will exhibit a hierarchical sequence
of crustal structure and epeirogenic behaviour. Studies of actual block-and-
basin mosaics confirm the presence of all these features, and indicate that
the all-important geometrical separation relationships are not only present
but exhibit a previously quite unsuspected degree of precision (a few
kilometres).

Chasmic basins within continents, even if they now lie near the
continental edge, are inherently sites where vast thicknesses of sediments
can have accumulated, the subsidence frequently occurring in two or more
distinct phases, owing to mechanical constraints. Many had restricted
marine circulation, and evaporites, usually much thicker than on
platforms, are common. Their hydrocarbon potential is further enhanced
by the opportunities for reservoir rock genesis and for trapping that are
offered by buried island inliers and at the steeply-bounded margins. It
appears probable, in fact, that many giant oil fields are located over buried
island inliers. Precise tectonic reconstructions can help in the precise
delineation of the original basin edges (i.e. the chasmic faults) and in the
detection and delineation of any buried inliers present. The large
dimensions of complete tectonic plates usually result in long separation
boundaries between any given pair of plates; consequently, precise tectonic
reconstruction studies at only two points along such an erstwhile boundary
will now be able to provide powerful exploration and interpretation
controls elsewhere along the boundary.

In their notable attempt at a plate tectonic analysis of the Alpide belt,
Dewey et al.[173] did not recognise that some of their inferred chasmic basins
might be of Palaeozoic age. Such an age, and probably pre-Devonian at
that, seems certain for most of the chasmic basins of north-western Europe.
In the United States, some appear to be even Late Precambrian in age. This
makes Late Precambrian and Palaeozoic plate tectonics an important
matter for understanding the basement control of hydrocarbon location in
these regions, and probably in many other parts of the world too. Thus it is
likely that, truly shield-based basins apart (e.g. Canadian, Russian, Persian

Gulf, Gangetic), chasmic basins of at least all Phanerozoic ages are the dominant control upon the present world distribution of hydrocarbon-prospective sedimentary bodies.

It appears that the interpretive framework outlined here can enable good progress to be made in the recognition, dating and precise delineation of these separative events, both in orogenic belts and elsewhere, offering valuable new insight into the development and structure of the corresponding basins. Particular mention in this regard may be made of the new interpretation of ophiolite emplacement, discussed briefly here in relation to the Caledonian belt and the Oman allochthon, which illustrates in an extreme form the risks of confusion between the phenomena of separative orogeny and those of approximative orogeny. For this reason the implications of the new plate model for plate approximation processes require much more thorough examination than the brief outline given here.

In conclusion, with the questions posed at the outset in mind, it would appear that, at least in the quest for hydrocarbons, plate tectonics is indeed capable of becoming a valuable, precise and pervasive basis for exploration and detailed interpretation.

Acknowledgements

In plate tectonics, perhaps more than in any other branch of Earth Science, one builds upon the ideas and data contributions of an enormous number of other people, and the foregoing account is no exception. The author hopes that this acknowledgement will, at least in part, make up for the inevitable limitations of the chosen reference list. This contribution benefited materially from comments by G. D. Hobson on an earlier manuscript. Adrina Aldridge gave much patient help with the preparation of the typescript.

REFERENCES

1. McKenzie, D. P. and Parker, R. L. (1967). The North Pacific: an example of tectonics on a sphere, *Nature*, **216**, 1276.
2. Morgan, W. J. (1968). Rises, trenches, great faults and crustal blocks, *J. Geophys. Res.*, **73**, 1959.
3. Le Pichon, X. (1968). Sea-floor spreading and continental drift, *J. Geophys. Res.*, **73**, 3661.
4. Heirtzler, J. R., Dickson, G. O., Herron, E. M., Pitman, W. C., III and Le Pichon, X. (1968). Marine magnetic anomalies, geomagnetic field reversals and motions of the ocean floor and continents, *J. Geophys. Res.*, **73**, 2119.

5. BELOUSOV, V. V. (1966). Modern concepts of the structure and development of the Earth's crust and the upper mantle of continents, *Quart. J. Geol. Soc. Lond.*, **122**, 293.
6. GREEN, D. H. and RINGWOOD, A. E. (1964). Fractionation of basalt magmas at high pressures, *Nature*, **201**, 1276.
7. GREEN, D. H. and RINGWOOD, A. E. (1967). The genesis of basaltic magmas, *Contrib. Miner. Petrol.*, **15**, 103.
8. GREEN, D. H. (1971). Composition of basaltic magmas as indicators of conditions of origin: application to oceanic volcanism, *Phil. Trans. Roy. Soc. Lond.*, **A268**, 707.
9. McKENZIE, D. P. and SCLATER, J. G. (1969). Heat flow in the Pacific and sea-floor spreading, *Bull. Volc.*, **33**, 101.
10. SCLATER, J. G. and FRANCHETEAU, J. (1970). The implications of terrestrial heat flow observations on current tectonic and geochemical models of the crust and upper mantle of the Earth, *Geophys. J. Roy. Astr. Soc.*, **20**, 509.
11. McKENZIE, D. P. and SCLATER, J. G. (1971). The evolution of the Indian Ocean since the Late Cretaceous, *Geophys. J. Roy. Astr. Soc.*, **24**, 437.
12. SCLATER, J. G., ANDERSON, R. N. and BELL, M. L. (1971). Elevation of ridges and evolution of the Central Eastern Pacific, *J. Geophys. Res.*, **76**, 7888.
13. McKENZIE, D. P. (1972). Plate tectonics, in: *The Nature of the Solid Earth* (E. C. Robertson, ed.) McGraw-Hill, New York, 323.
14. TRÉHU, A. M. (1975). Depth versus (age)$^{1/2}$: a perspective on mid-ocean rises, *Earth Planet. Sci. Letters*, **27**, 287.
15. OLIVER, J. and ISACKS, B. (1967). Deep earthquake zones, anomalous structures in the upper mantle, and the lithosphere, *J. Geophys. Res.*, **72**, 4259.
16. ISACKS, B., OLIVER, J. and SYKES, L. R. (1968). Seismology and the new global tectonics, *J. Geophys. Res.*, **73**, 5855.
17. TURCOTTE, D. L. and OXBURGH, E. R. (1972). Mantle convection and the new global tectonics, *Ann. Rev. Fluid. Mech.*, **4**, 33.
18. BERNAL, J. D. (1961). Continental and oceanic differentiation, *Nature*, **192**, 123.
19. DORMAN, J., EWING, M. and OLIVER, J. (1960). Study of the shear velocity distribution in the upper mantle by mantle Rayleigh waves, *Bull. Seismol. Soc. Am.*, **50**, 87.
20. MACDONALD, G. J. F. (1963). The deep structure of continents, *Rev. Geophys.*, **1**, 587.
21. MACDONALD, G. J. F. (1965). Continental structure and drift, in: *A Symposium on Continental Drift* (P. M. S. Blackett, E. Bullard and S. K. Runcorn, eds.) Phil. Trans. Roy. Soc. Lond. No. 1088, 215.
22. ANDERSON, D. L. (1967). Latest information from seismic observations, in: *The Earth's Mantle* (T. F. Gaskell, ed.) Academic Press, London, 355.
23. KNOPOFF, L. (1969). The upper mantle of the Earth, *Science*, **163**, 1277.
24. JORDAN, T. H. (1975). The continental tectosphere, *Rev. Geophys. Space Phys.*, **13**, 1.
25. JORDAN, T. H. and FYFE, W. S. (1976). Penrose Conference Report: Lithosphere–asthenosphere boundary, *Geology*, **4**, 770.
26. BROOKS, C., JAMES, D. E. and HART, S. R. (1976). Ancient lithosphere: its role in young continental volcanism, *Science*, **193**, 1086.

27. MORGAN, W. J. (1975). Heat flow and vertical movements of the crust, in: *Petroleum and Global Tectonics* (A. G. Fischer and S. Judson, eds.) Princeton Univ. Press, 23.

28. ANDERSON, D. L. and SPETZLER, H. (1970). Partial melting and the low-velocity zone, *Phys. Earth Planet. Interiors*, **4**, 62.

29. ANDERSON, D. L. (1970). Petrology of the mantle, *Mineral. Soc.*, Spec. Paper 3, 85.

30. BIRCH, F. (1969). Density and composition of the upper mantle: first approximation as an olivine layer, in: *The Earth's Crust and Upper Mantle* (P. J. Hart, ed.) Am. Geophys. Union Geophys. Monogr. 13, 18.

31. LAMBERT, I. B. and WYLLIE, P. J. (1970). Low velocity zone of the Earth's mantle: incipient melting caused by water, *Science*, **169**, 764.

32. LEEDS, A. R., KNOPOFF, L. and KAUSEL, E. G. (1974). Variations of upper mantle structure under the Pacific Ocean, *Science*, **186**, 141.

33. FORSYTH, D. W. (1975). The early structural evolution and anisotropy of the oceanic upper mantle, *Geophys. J. Roy. Astr. Soc.*, **43**, 103.

34. EGGLER, D. H. (1976). Does CO_2 cause partial melting in the low-velocity layer of the mantle? *Geology*, **4**, 69.

35. SCLATER, J. G., LAWVER, L. A. and PARSONS, B. (1975). Comparison of long-wavelength residual elevation and free air gravity anomalies in the North Atlantic and possible implications for the thickness of the lithosphere plate, *J. Geophys. Res.*, **80**, 1031.

36. OSMASTON, M. F. (1973). Limited lithosphere separation as a main cause of continental basins, continental growth and epeirogeny, in: *Implications of Continental Drift to the Earth Sciences*, 2 (D. H. Tarling and S. K. Runcorn, eds.) Academic Press, London, 649.

37. MORGAN, W. J. (1972). Deep mantle convection plumes and plate motions, *Am. Assoc. Petrol. Geologists Bull.*, **56**, 203.

38. BAILEY, D. K. (1977). Lithosphere control of continental rift magmatism, *J. Geol. Soc. Lond.*, **133**, 103.

39. O'HARA, M. J. (1965). Primary magmas and the origin of basalts, *Scottish J. Geol.*, **1**, 19.

40. O'HARA, M. J. (1970). Upper mantle composition inferred from laboratory experiments and observation of volcanic products, *Phys. Earth Planet. Interiors*, **3**, 236.

41. RINGWOOD, A. E. and LOVERING, J. F. (1970). Significance of pyroxene–ilmenite intergrowths among kimberlite xenoliths, *Earth Planet. Sci. Letters*, **7**, 371.

42. DAWSON, J. B. (1972). Kimberlites and their relation to the mantle, *Phil. Trans. Roy. Soc. Lond.*, **A271**, 297.

43. BOYD, F. R. and NIXON, P. H. (1973). Origin of the ilmenite–silicate nodules in kimberlites from Lesotho and South Africa, in: *Lesotho Kimberlites* (P. H. Nixon, ed.) Lesotho National Development Corporation, Cape Town, 254.

44. OSMASTON, M. F. (1971). Genesis of ocean ridge median valleys and continental rift valleys, *Tectonophysics*, **11**, 387.

45. BURKE, K. and WHITEMAN, A. J. (1973). Uplift, rifting and the break-up of Africa, in: *Implications of Continental Drift to the Earth Sciences*, 2 (D. H. Tarling and S. K. Runcorn, eds.) Academic Press, London, 735.

46. BLUNDELL, D. J., DAVEY, F. J. and GRAVES, L. J. (1971). Geophysical surveys over the South Irish Sea and Nymphe Bank, *J. Geol. Soc. Lond.*, **127**, 339.
47. DOBSON, M. R., EVANS, W. E. and WHITTINGTON, R. (1973). The geology of the South Irish Sea, *Inst. Geol. Sci.*, Report 73/11.
48. HALL, J. and SMYTHE, D. K. (1973). Discussion of the relation of Palaeogene ridge and basin structures of Britain to the North Atlantic, *Earth Planet. Sci. Letters*, **19**, 54.
49. ROBERTS, D. G. (1974). Structural development of the British Isles, the continental margin and the Rockall Plateau, in: Ref. 106, 343.
50. LAUGHTON, A. S. (1975). Tectonic evolution of the northeast Atlantic Ocean: a review, *Norges. Geol. Unders.*, **316**, 169.
51. WOODLAND, A. W. (ed.) (1975). *Petroleum and the Continental Shelf of Northwest Europe*, **1**, Applied Science Publishers, Barking, England.
52. PARKIN, L. W. (ed.) (1969). *Handbook of South Australian Geology*, Geol. Surv. South Aust., Adelaide.
53. KING, L. C. (1967). *Morphology of the Earth*, 2nd edn., Oliver and Boyd, Edinburgh.
54. OLIVER, J., ISACKS, B. L. and BARAZANGI, M., Seismicity at continental margins, in: Ref. 106, 85.
55. KENNETT, J. P., HOUTZ, R. E., *et al.* (eds.) (1974). *Initial reports of the Deep Sea Drilling Project*, **29**, Washington (US Govt. Printing Office), Site 283.
56. CAMPOS, C. W. M., PONTE, F. C. and MIURA, K. (1974). Geology of the Brazilian continental margin, in: Ref. 106, 447.
57. MOORE, A. E. (1976). Controls of post-Gondwanaland alkaline volcanism in southern Africa, *Earth Planet. Sci. Letters*, **31**, 291.
58. LANGSETH, M. G., LE PICHON, X. and EWING, M. (1966). Crustal structure of the mid-ocean ridges. 5. Heat flow through the Atlantic Ocean floor and convection currents, *J. Geophys. Res.*, **71**, 5321.
59. LAMBECK, K. (1972). Gravity anomalies over ocean ridges, *Geophys. J. Roy. Astr. Soc.*, **30**, 37.
60. LE PICHON, X., FRANCHETEAU, J. and BONNIN, J. (1973). *Plate Tectonics*, Elsevier, Amsterdam.
61. FISCHER, A. G. (1975). Origin and growth of basins, in: *Petroleum and Global Tectonics* (A. G. Fischer and S. Judson, eds.) Princeton Univ. Press, 47.
62. WOOLLARD, G. P. (1968). The interrelationship of the crust, the upper mantle and isostatic gravity anomalies in the United States, in: *The Crust and Upper Mantle of the Pacific Area* (L. Knopoff, C. L. Drake and P. J. Hart, eds.) Am. Geophys. Union Geophys. Monogr. 12, 312.
63. QURESHY, M. N. (1971). Relation of gravity to elevation and rejuvenation of blocks in India, *J. Geophys. Res.*, **76**, 545.
64. TURCOTTE, D. L. (1974). Are transform faults thermal contraction cracks? *J. Geophys. Res.*, **79**, 2573.
65. COLLETTE, B. J. (1974). Thermal contraction joints in a spreading sea-floor as origin of fracture zones, *Nature*, **251**, 299.
66. MCKENZIE, D. P. (1967). Some remarks on heat flow and gravity anomalies, *J. Geophys. Res.*, **72**, 6261.
67. TALWANI, M. and LE PICHON, X. (1969). Gravity field over the Atlantic ocean,

in: *The Earth's Crust and Upper Mantle* (P. J. Hart, ed.) Am. Geophys. Union Geophys. Monogr. 13, 341.

68. KAULA, W. M. (1970). Earth's gravity field: relation to global tectonics, *Science*, **169**, 982.

69. KAULA, W. M. (1972). Global gravity and tectonics, in: *The Nature of the Solid Earth* (E. C. Robertson, ed.) McGraw-Hill, New York, 385.

70. ANDERSON, R. N., MCKENZIE, D. and SCLATER, J. G. (1973). Gravity, bathymetry and convection in the Earth, *Earth Planet. Sci. Letters*, **18**, 391.

71. NAFE, J. E. and DRAKE, C. L. (1969). Floor of the North Atlantic—summary of geophysical data, in: *North Atlantic—Geology and Continental Drift* (M. Kay, ed.) Am. Assoc. Petrol. Geologists Mem. 12, 59.

72. RONA, P. A. (1973). Relations between rates of sediment accumulation on continental shelves, sea-floor spreading and eustasy inferred from the central North Atlantic, *Geol. Soc. Am. Bull.*, **84**, 2851.

73. DIETZ, R. S. (1963). Collapsing continental rises: an actualistic concept of geosynclines and mountain building, *J. Geol.*, **71**, 314.

74. DIETZ, R. S. and HOLDEN, J. C. (1974). Collapsing continental rises: actualistic concept of geosynclines—a review, in: Ref. 163, 14.

75. SLEEP, N. H. (1973). Crustal thinning on Atlantic continental margins; evidence from older margins, in: *Implications of Continental Drift to the Earth Sciences*, **2** (D. H. Tarling and S. K. Runcorn, eds.) Academic Press, London, 685.

76. BOTT, M. H. P. (1971). Evolution of young continental margins and formation of shelf basins, *Tectonophysics*, **11**, 319.

77. BOTT, M. H. P. and DEAN, D. S. (1972). Stress systems at young continental margins, *Nature Phys. Sci.*, **235**, 23.

78. BOTT, M. H. P. (1973). Shelf subsidence in relation to the evolution of young continental margins, in: *Implications of Continental Drift to the Earth Sciences* **2** (D. H. Tarling and S. K. Runcorn, eds.) Academic Press, London, 675.

79. KENT, P. E. (1975). Review of the North Sea basin development, *J. Geol. Soc. Lond.*, **131**, 435.

80. KENT, P. E. (1975). The tectonic development of Great Britain and the surrounding seas, in: Ref. 51, 3.

81. ZIEGLER, P. A. (1975). North Sea basin history in the tectonic framework of north-western Europe, in: Ref. 51, 131.

82. ZIEGLER, W. H. (1975). Outline of the geological history of the North Sea, in: Ref. 51, 165.

83. WHITEMAN, A. J., REES, G., NAYLOR, D. and PEGRUM, R. P. (1975). North Sea troughs and plate tectonics, in: *Petroleum Geology and Geology of the North Sea and North-east Atlantic Margin* (A. J. Whiteman, D. Roberts and M. A. Sellevoll, eds.) Norges. Geol. Unders. **316**, 137.

84. ROBERTS, D. G. (1975). Marine geology of the Rockall Plateau and Trough, *Phil. Trans. Roy. Soc. Lond.*, **A278**, 447.

85. ROBERTS, D. G. (1975). Tectonic and stratigraphic evolution of the Rockall Plateau and Trough, in: Ref. 51, 77.

86. HOWITT, F., ASTON, E. R. and JACQUE, M. (1975). Occurrence of Jurassic volcanics in the North Sea, in: Ref. 51, 379.

87. STRIDE, A. H. CURRAY, J. R., MOORE, D. G. and BELDERSON, R. H. (1969). Marine geology of the Atlantic continental margin of Europe, *Phil. Trans. Roy. Soc. Lond.*, **A264**, 31.
88. CLARKE, R. H., BAILEY, R. J. and TAYLOR-SMITH, D. (1971). Seismic reflection profiles of the continental margin west of Ireland, in: *The Geology of the East Atlantic Continental Margin*. 2. *Europe* (F. M. Delany, ed.) Inst. Geol. Sci. Report 70/14, 67.
89. HALLAM, A. and SELLWOOD, B. W. (1976). Middle Mesozoic sedimentation in relation to tectonics in the British area, *J. Geol.*, **84**, 301.
90. SHERIDAN, R. E. (1974). Atlantic continental margin of North America, in: Ref. 106, 391.
91. OSMASTON, M. F. (1977). Origin and emplacement of ophiolites: Discussion of a paper by I. G. Gass, in: *Volcanic Processes in Ore Genesis*, Geol. Soc. Lond. Spec. Publ. 7, 171.
92. FALCON, N. L. and KENT, P. E. (1960). Geological results of petroleum exploration in Britain 1945–1957, *Geol. Soc. Lond. Mem.* 2.
93. LARSON, P. A., MUDIE, J. D. and LARSON, R. L. (1972). Magnetic anomalies and fracture-zone trends in the Gulf of California, *Geol. Soc. Am. Bull.*, **83**, 3361.
94. DIETZ, R. S. and SPROLL, W. P. (1970). East Canary Islands as a microcontinent within Africa–North America continental drift fit, *Nature*, **226**, 1043.
95. BECK, R. H. and LEHNER, P. (1974). Oceans, new frontier in exploration, *Am. Assoc. Petrol. Geologists Bull.*, **58**, 376.
96. STILLMAN, C. J., FUSTER, J. M., BENNELL-BAKER, M. J., MUNOZ, M., SMEWING, J. D. and SAGREDO, N. (1975). Basal complex of Fuerteventura (Canary Islands) is an oceanic intrusive complex with rift-system affinities, *Nature*, **257**, 469.
97. CAREY, S. W. (1958). The tectonic approach to continental drift in: *Continental Drift: A Symposium* 1956 (S. W. Carey, ed.) Univ. of Tasmania, Hobart, 177.
98. VAN ANDEL, S. I. and HOSPERS, J. (1970). A critical study of the rhombochasm concept as applied to the West Siberian lowlands, in: *Palaeogeophysics* (S. K. Runcorn, ed.) Academic Press, London, 313.
99. BAZANOV, E. A. *et al.* (1967). Gas and oil-bearing provinces of Siberia, *Proc. 7th World Petroleum Congr.* **2**, 109.
100. EMERY, K. O., UCHUPI, E., BOWIN, C. O., PHILLIPS, J. and SIMPSON, E. S. W. (1975). Continental margin off south-west Africa, Cape St. Francis (South Africa) to Walvis Ridge, *Am. Assoc. Petrol. Geologists Bull.*, **59**, 3.
101. EMERY, K. O., UCHUPI, E., PHILLIPS, J. D., BROWN, C. and MASCLE, J. (1975). Continental margin off western Africa, *Am. Assoc. Petrol. Geologists Bull.*, **59**, 2209.
102. RIGASSI, D. A. (1975). Petroleum geology of Gabon basin: discussion, *Am. Assoc. Petrol. Geologists Bull.*, **59**, 542.
103. ILLIES, J. H., and MUELLER, ST. (eds.) (1970). *Graben problems*, Schweizerbart, Stuttgart; International Upper Mantle Project, Scientific Rep. 27.
104. FLORENSOV, N. A. (1969). Rifts of the Baikal mountain region, *Tectonophysics*, **8**, 443.

105. BAKER, B. H., MOHR, P. A. and WILLIAMS, L. A. J. (1972). Geology of the eastern rift system of Africa, *Geol. Soc. Am.*, Spec. Paper 136.
106. BURK, C. A. and DRAKE, C. L. (eds.) (1974). *The Geology of Continental Margins*, Springer-Verlag, Berlin.
107. LLIBOUTRY, L. (1969). Sea-floor spreading, continental drift and lithosphere sinking with an asthenosphere at melting point, *J. Geophys. Res.*, **74**, 6525.
108. HATHERTON, T. (1974). Active continental margins and island arcs, in: Ref. 106, 93.
109. SEELY, D. R., VAIL, P. R. and WALTON, G. G. (1974). Trench slope model, in: Ref. 106, 249.
110. MITCHELL, A. H. and READING, H. G. (1971). Evolution of island arcs, *J. Geol.*, **79**, 253.
111. STEWART, J. W., EVERNDEN, J. F. and SNELLING, N. J. (1974). Age determinations from Andean Peru: a reconnaissance survey, *Geol. Soc. Am. Bull.*, **85**, 1107.
112. FARRAR, F., CLARK, A. H., HAYNES, S. J., QUIRT, G. S., COWN, H. and ZENTILLI, M. (1971). K–Ar evidence for the post-Paleozoic migration of the granitic intrusions foci in the Andes of northern Chile, *Earth Planet. Sci. Letters*, **10**, 60.
113. VERGARA, M. and MUNIZAGA, F. (1974). Age and evolution of the Upper Cenozoic andesitic volcanism in Central–South Chile, *Geol. Soc. Am. Bull.*, **85**, 603.
114. KARIG, D. (1974). Tectonic erosion at trenches, *Earth Planet. Sci.Letters*, **21**, 209.
115. WHITFORD, D. J. (1975). Strontium isotopic studies of the volcanic rocks of the Sunda arc, Indonesia, and their petrogenetic implications, *Geochim. Cosmochim. Acta*, **39**, 1287.
116. PACKHAM, G. H. and FALVEY, D. A. (1971). An hypothesis for the formation of marginal seas in the western Pacific, *Tectonophysics*, **11**, 79.
117. UYEDA, S. (1974). North-west Pacific trench margins, in: Ref. 106, 473.
118. MINATO, M., GORAI, M. and HUNAHASHI, M. (eds.) (1965). *The geological development of the Japanese Islands*, Tsukiji Shokan, Tokyo.
119. MATSUDA, T. and UYEDA, S. (1971). On the Pacific-type orogeny and its model—extension of the paired belts concept and possible origin of marginal seas, *Tectonophysics*, **11**, 5.
120. KARIG, D. E. (1971). Origin and development of marginal basins in the western Pacific, *J. Geophys. Res.*, **76**, 2542.
121. HULL, C. E. and WARMAN, H. R. (1970). Asmari oil fields of Iran, in: *Geology of Giant Petroleum Fields* (M. T. Halbouty, ed.) Am. Assoc. Petrol. Geologists, Mem. 14, 428.
122. DEWEY, J. F. and BIRD, J. M. (1970). Mountain belts and the new global tectonics, *J. Geophys. Res.*, **75**, 2625.
123. LE FORT, P. (1975). Himalayas: the collided range. Present knowledge of the continental arc, in: *Tectonics and Mountain Ranges* (J. H. Ostrom and P. M. Orville, eds.) *Am. J. Sci.*, **275A**, 1.
124. AUBOUIN, J. (1965). *Geosynclines*, Elsevier, Amsterdam.
125. WILSON, H. H. (1969). Late Cretaceous eugeosynclinal sedimentation,

gravity tectonics, and ophiolite emplacement in Oman Mountains, south-east Arabia, *Am. Assoc. Petrol. Geologists Bull.*, **53**, 626.

126. CHURCH, W. R. and STEVENS, R. K. (1971). Early Paleozoic ophiolite complexes of the Newfoundland Appalachians as mantle-oceanic crust sequences, *J. Geophys. Res.*, **76**, 1460.

127. CHURCH, W. R. (1972). Ophiolite: its definition, origin as oceanic crust, and mode of emplacement in orogenic belts, with special reference to the Appalachians, in, *The Ancient Oceanic Lithosphere*, Earth Phys. Branch, Dept. Energy Mines and Resources, Canada, **42**(3), 71.

128. WILLIAMS, H. and SMYTH, W. R. (1973). Metamorphic aureoles beneath ophiolite suites and Alpine peridotites, *Am. J. Sci.*, **273**, 594.

129. MALPAS, J. G. (in press). Magma generation in the upper mantle, field evidence from ophiolite suites, and application to the generation of oceanic lithosphere, in: *Proc. Disc. Meet. 'Terrestrial heat and the generation of magmas'* (G. M. Brown, M. J. O'Hara and E. R. Oxburgh, organisers) Phil. Trans. Roy. Soc. Lond. Ser. A.

130. DEWEY, J. F. (1974). Continental margins and ophiolite obduction: Appalachian Caledonian system, in: Ref. 106, 933.

131. GLENNIE, K. W., BOEUF, M. G. A., HUGHES CLARKE, M. W., MOODY-STUART, M., PILAAR, W. F. H. and REINHARDT, B. M. (1973). Late Cretaceous nappes in Oman Mountains and their geologic evolution, *Am. Assoc. Petrol. Geologists Bull.*, **57**, 5.

132. STONELEY, R. (1974). Evolution of the continental margins bounding a former southern Tethys, in Ref. 106, 889.

133. HESS, H. H. (1939). Island arcs, gravity anomalies and serpentine intrusions: a contribution to the ophiolite problem, *Proc. 17th Internat. Geol. Congr.* (Moscow 1937) **2**, 263.

134. COBBING, E. J. (1972). Tectonic elements of Peru and the evolution of the Andes, *Proc. 24th Internat. Geol. Congr.* (Montreal) **3**, 306.

135. SUGIMURA, A. and UYEDA, S. (1973). *Island Arcs: Japan and its Environs*, Elsevier, Amsterdam.

136. DEWEY, J. F. (1969). Evolution of the Appalachian/Caledonian orogen, *Nature*, **222**, 124.

137. FITTON, J. G. and HUGHES, D. J. (1970). Volcanism and plate tectonics in the British Ordovician, *Earth Planet. Sci. Letters*, **8**, 223.

138. STILLMAN, C. J., DOWNES, K. and SCHEINER, E. J. (1974). Caradocian volcanic activity in east and south-east Ireland, *Sci. Proc. Roy. Dublin Soc.*, **5**, 87.

139. MERCY, E. L. P. (1965). Caledonian igneous activity, in: *The Geology of Scotland* (G. Y. Craig, ed.) Oliver and Boyd, Edinburgh, 229.

140. WILLIAMS, H. and STEVENS, R. K. (1974). The ancient continental margin of eastern North America, in: Ref. 106, 781.

141. LAURENT, R. (1975). Occurrences and origin of the ophiolites of southern Quebec, Northern Appalachians, *Can. J. Earth Sci.*, **12**, 443.

142. STÖCKLIN, J, (1974). Possible ancient continental margins in Iran, in: Ref. 106, 873.

143. CRAWFORD, A. R. (1973). A displaced Tibetan massif as a possible source of some Malayan rocks, *Geol. Mag.*, **109**, 483.

144. GANSSER, A. (1964). *Geology of the Himalayas*, Wiley-Interscience, New York.
145. REINHARDT, B. M. (1969). On the genesis and emplacement of ophiolites in the Oman Mountains geosyncline, *Schweiz. Min. Pet. Mitt.*, **49,** 1.
146. GANSSER, A. (1973). *Himalaya*, in: *Mesozoic-Cenozoic Orogenic Belts* (A. M. Spencer, ed.) Geol. Soc. Lond. Spec. Publ. 4, 267.
147. FALCON, N. L. (1958). Position of oil fields of south-west Iran with respect to relevant sedimentary basins, in: *Habitat of Oil* (L. G. Weeks, ed.) Am. Assoc. Petrol. Geologists, 1279.
148. WELLS, A. J. (1969). The crush zone of the Iranian Zagros Mountains and its implications, *Geol. Mag.*, **106**, 385.
149. KAMEN-KAYE, M. (1970). Geology and productivity of the Persian Gulf Synclinorium, *Am. Assoc. Petrol. Geologists Bull.*, **54**, 2371.
150. FALCON, N. L. (1973). Southern Iran: Zagros Mountains, in: *Mesozoic–Cenozoic Orogenic Belts* (A. M. Spencer, ed.) Geol. Soc. Lond. Spec. Publ. 4, 199.
151. HAYNES, S. J. and McQUILLAN, H. (1974). Evolution of the Zagros suture zone, southern Iran, *Geol. Soc. Am. Bull.*, **85**, 739.
152. TIRATSOO, E. N. (1976). *Oilfields of the World*, 2nd edn., Scientific Press, Beaconsfield.
153. COLEMAN, R. G. (1974). Geologic background of the Red Sea, in: Ref. 106, 743.
154. WHITE, R. S. and KLITGORD, K. (1976) Sediment deformation and plate tectonics in the Gulf of Oman, *Earth Planet. Sci. Letters*, **32**, 199.
155. KING, B. C. (1976). The Baikal Rift, *J. Geol. Soc. Lond.*, **132**, 348.
156. BLAIR, D. G. (1975). Structural styles in North Sea oil and gas fields, in: Ref. 51, 327.
157. MITCHELL, A. H. and READING, H. G. (1969). Continental margins, geosynclines, and ocean floor spreading, *J. Geol.*, **77**, 629.
158. HEDBERG, H. D. (1970). Continental margins from viewpoint of the petroleum geologist, *Am. Assoc. Petrol. Geologists Bull.*, **54**, 3.
159. READING, H. G. (1972). Global tectonics and the genesis of flysch successions, *Proc.* 24th Internat. Geol. Congr. (Montreal) **6**, 59.
160. KINSMAN, D. J. (1975). Rift valley basins and sedimentary history of trailing continental margins, in: *Petroleum and Global Tectonics* (A. G. Fischer and S. Judson, eds.) Princeton Univ. Press, 83.
161. DICKINSON, W. R. (1974). Plate tectonics and sedimentation, in: *Tectonics and Sedimentation* (W. R. Dickinson, ed.) Soc. Econ. Pal. Min. Spec. Publ. 22, 1.
162. DOTT, R. H., JR. (1974). The geosynclinal concept, in: Ref. 163, 1.
163. DOTT, R. H., JR. and SHAVER, R. H. (eds.) (1974). *Modern and ancient geosynclinal sedimentation*, Soc. Econ. Pal. Min., Spec. Publ. 19.
164. PARKER, J. R. (1975). Lower Tertiary sand development in the central North Sea, in: Ref. 51, 447.
165. KAY, M. (1974). Reflections: geosynclines, flysch and melanges, in: Ref. 163, 377.
166. HALBOUTY, M. T., KING, R. E., KLEMME, H. D., DOTT, R. H., SR. and MEYERHOFF, A. A. (1970). Factors affecting formation of giant oil and gas fields, and basin classification, Am. Assoc. Petrol. Geologists Mem. 14, 528.

167. DETRICK, R. S., SCLATER, J. G. and THIEDE, J. (1977). The subsidence of aseismic ridges, *Earth Planet. Sci. Letters*, **34**, 185.
168. KLEMME, H. D. (1975). Geothermal gradients, heat flow and hydrocarbon recovery, in: *Petroleum and Global Tectonics* (A. G. Fischer and S. Judson, eds.) Princeton Univ. Press, 251.
169. GATEWOOD, L. E. (1970). Oklahoma City field—anatomy of a giant, in:*Geology of Giant Petroleum Fields* (M. T. Halbouty, ed.) Am. Assoc. Petrol. Geologists, Mem. 14, 223.
170. HAM, W. E. and WILSON, J. L. (1967). Paleozoic epeirogeny and orogeny in the central United States, *Am. J. Sci.*, **265**, 332.
171. HOFFMAN, P., DEWEY, J. F. and BURKE, K. (1974). Aulacogens and their genetic relation to geosynclines, with a Proterozoic example from Great Slave Lake, Canada, in: Ref. 163, 38.
172. VINIEGRA O. F. and CASTILLO-TEJERO, C. (1970). Golden Lane fields, Veracruz, Mexico, in: *Geology of Giant Petroleum Fields* (M. T. Halbouty, ed.) Am. Assoc. Petrol. Geologists Mem. 14, 309.
173. DEWEY, J. F., PITMAN, W. C., III, RYAN, W. B. F. and BONNIN, J. (1973). Plate tectonics and the evolution of the Alpine System, *Geol. Soc. Am. Bull.*, **84**, 3137.
174. HOLT, J. B. (1975). Thermal diffusivity of olivine, *Earth Planet. Sci.Letters*, **27**, 404.

Note added in proof. Arguments that major ophiolite allochthons are emplaced onto continental margins during the early stages of plate separation, set out on pp. 24 and 31, may suggest a cavalier disregard for the widely held view that most ophiolites have, on the contrary, been emplaced wholly by subduction-related processes. In fact, while the author readily accepts that belts of crustal rocks, tectonically accreted and metamorphosed during subduction (see p. 29), might incorporate sheared-off ophiolitic portions of oceanic igneous basement, a satisfactory example of this seems hard to find. For instance, in the classic western California example of such a belt (see several papers in ref. 106) this type of interpretation simply does not fit the facts. There, the main ophiolite belt forms the Tithonian base of the non-accreted Great Valley Sequence, and seems best interpreted as having been shelf-emplaced during a Tithonian separative rupture along a line just within the original continental margin. The like-aged ophiolite blocks in the blueschist-metamorphosed Franciscan terrain, to the west, could have been emplaced at this time onto the corresponding western edge of the new chasmic basin, followed by deformation and metamorphism when that basin closed again (with longitudinal offset?) during the Cretaceous.

Unfortunately, a major paper on ophiolite emplacement by M. E. Brookfield (*Tectonophysics* (1977) **34**, 247), appeared far too late for discussion in the text. He, too, concludes that subduction-related obduction is irrelevant.

Finally, it has been pointed out to the author by Sir Peter Kent that the Mesozoic stratigraphy of the Iranian Lut-Nain block does not support the proposal (p. 33) that the area supplied the Hawasina pile of nappes in Oman. Instead, the Hawasina rocks appear (ref. 131) to have floored and margined a rather wide chasmic basin of Permian or Triassic age, which must have lain between Arabia and Iran but now has largely disappeared by subduction. The Semail ophiolite nappe *could* have been generated and emplaced by the Late Cretaceous rifting within this basin, but the proposal in the text that it is far-travelled from the north better explains the wide distribution of ophiolite mélanges in Iran.

Chapter 2

THE APPLICATION OF THE RESULTS OF ORGANIC GEOCHEMICAL STUDIES IN OIL AND GAS EXPLORATION

B. Tissot

Institut Français du Pétrole, France

SUMMARY

Recent advances in organic geochemistry now offer a new approach to petroleum exploration. The application of the principles of petroleum generation and migration makes possible a systematic appraisal of the petroleum potential of a sedimentary basin. Identification of source rocks by quick methods, and quantitative evaluation of their petroleum potential, can be done either in the laboratory or on the well-site. Oil/source rock and oil/oil correlations complement the definition of the plays and targets present in a sedimentary basin. Optical and chemical indices contribute to the definition of the prospects, together with the use of mathematical models. Timing of petroleum generation can now be evaluated and compared with the age of the formation of traps.

INTRODUCTION

The aim of applying organic geochemistry to petroleum exploration is to define the various targets, which may occur in different geological horizons, and to locate the most favourable areas for exploring those targets. To achieve this objective, cores, cuttings and crude oils or gas samples from exploration wells, and rocks from outcrops are analysed.

The main problems to be solved are the evaluation of the source rocks, oil/oil and oil/source rock correlation, and the timing of hydrocarbon generation.

SOURCE ROCK EVALUATION

Three conditions have to be realised in order to generate petroleum or gas from a possible source rock:

—A sufficient amount of organic matter.

—The right quality of organic matter: the chemical composition of kerogen should be favourable for a high yield of oil and gas upon burial. This parameter may be quantified and expressed by the *genetic potential*, i.e. the total amount of oil and gas that the source rock is able to generate, if it is buried at a sufficient depth for long enough.

—The maturation of the source rock: the thermal history of the source rock should be such as to produce a significant part of the petroleum which could be expected from the nature of the organic matter. The maturation may be expressed by the *transformation ratio*, i.e. the ratio of the petroleum already generated to the genetic potential.

In addition, a proper source rock evaluation ideally should also provide an indicator of migration out of the source rock. However, such an index is not available with the present state of knowledge on the mechanisms of petroleum expulsion from the source rock to the reservoir rock. Furthermore, the most recent data on this subject suggest that the more hydrocarbons are produced, generating high pressure in a source rock, the more favourable are the conditions for expulsion. Thus, a high transformation ratio would be also a favourable indication for migration.

The Abundance of Organic Matter

This is determined by measurement of the organic carbon content of the rock. The analysis is performed on ground rock, which may or may not previously have been extracted by solvent. Carbonates are first destroyed by hydrochloric acid treatment in order to eliminate them as potential sources of carbon dioxide. Then, the organic matter is burnt in an oxygen atmosphere; induction furnaces are particularly convenient for this purpose, since they reach a high temperature in a very short time. The amount of carbon dioxide generated by combustion is measured and thus provides the carbon content expressed as percent of the rock.

It is generally considered that a carbon content of 0.5% represents a lower limit for an effective source rock. In fact, whatever the nature of the organic matter, an original carbon content lower than 0.5% is likely to produce insufficient amounts of liquid petroleum with respect to the adsorption properties of the source rock and to the necessary pressure build-up for the expulsion of oil. However, in the deeper parts of sedimentary basins, where the maturation of the organic matter has reached the dry gas stage, a residual 0·5 % organic carbon may be witness of an original 1 % or more. Thus, it may have been a potential source rock for gas.

Type, Quality and Maturation Stage of the Organic Matter

The type and quality of the organic matter on the one hand, and its maturation stage on the other can be assessed by several techniques: chemical or physical analysis of kerogen and pyrolysis performed on the whole rock; chemical analysis of bitumen recovered by solvent extraction, and optical examination of the kerogen. Each of these procedures has its own advantages and preferred field of application. However, the major differences are between optical and physico–chemical techniques. Optical methods allow visualisation of the organic matter and thus an appreciation of the mixing of organic sources, either contemporaneous—e.g. marine plankton and continental detritus—or reworked from older rocks. However, most optical techniques are concerned with only a minor part of the organic matter which is not necessarily representative of the total; e.g. relatively large fragments or concentrates obtained by difference of density, wettability, etc. In contrast, physico–chemical techniques involve the whole of the organic matter, and the analysis can be considered as representative. However, mixing or reworking of the various organic materials is often difficult to detect or to evaluate quantitatively. Therefore, it seems advisable to associate, as far as is practicable, an optical and a physico–chemical technique. This combination is able to provide a correct answer to the questions of type and maturation of the organic matter in almost all cases.

Physico–Chemical Analysis of Kerogen

This is the basis for fundamental studies on the characterisation of the types of organic matter.[25] It is also one of the most reliable ways for characterising the stage of maturation; in particular, it is unaffected by migration and accumulation phenomena, which may increase the content of soluble organic matter and alter the composition of the extract. Kerogen analyses, however, require preliminary isolation of the kerogen by hydrochloric and hydrofluoric destruction of the inorganic minerals. The efficiency of the elimination of such minerals is measured by the ash content of the residual kerogen. The main impurity remaining after the acid treatment is pyrite. When the pyrite content is high, it is necessary to attempt separation by using the difference in specific gravity between organic matter and pyrite. However, the pyrite–organic matter association is sometimes very intimate. For example, pyrite micro-crystals may be coated with organic matter and assembled to form raspberry-like aggregates (framboids). In such a case, it is obviously not possible to achieve separation and a certain percentage of pyrite has to be accepted; otherwise, there would be a loss of kerogen.

Elemental analysis of kerogen based on the major atomic constituents C, H and O has been used to define three main types of kerogen.

Type I refers to kerogens having originally a high hydrogen content and a low oxygen content. This type of kerogen is very rich in lipids, and particularly aliphatic chains. It contains only a small proportion of polyaromatic nuclei and oxygenated functional groups. The organic matter may be derived from an accumulation of algae, like oil shales containing Botryococcus and the Recent coorongite. It may also result from an intense reworking of various kinds of organic matter by microbes, leaving mostly the lipid fraction of the original material, and microbial lipids. The well known Green River shales of Colorado and Utah belong to this type of organic sedimentation. Type I kerogen has a high genetic potential for petroleum. It also comprises the richest oil shales.

Type II refers to kerogen having a fairly high original hydrogen content, although it is somewhat lower than in Type I. Kerogen contains abundant aliphatic chains of various lengths and saturated rings. Also, it comprises polyaromatic nuclei and hetero-atomic functional groups, like carbonyl or carboxyl groups, which contain oxygen. The organic matter is usually derived from marine phytoplankton and zooplankton, laid down in a confined environment. Many classical oil source rocks belong to this type: Silurian of North Africa, Cretaceous of the Middle East, Jurassic of Western Europe, etc. Some oil shales, like kukersite of the USSR or the Irati shales of Brazil, also contain Type II kerogen.

Type III refers to kerogen having a low hydrogen content and a high original oxygen content. Its chemical structure comprises mainly polyaromatic nuclei and oxygen-containing functional groups. In addition to these some aliphatic chains, including long straight chains from natural waxes, are attached to the polycyclic network. The organic matter is mostly derived from the higher land plants. This terrestrial input may be transported by rivers and currents and laid down either in non-marine or in deltaic or continental margin environments. Type III kerogen has a comparatively low potential for oil, but it can provide a convenient gas source at depth. The lower Mannville shales of western Canada, or the Upper

Cretaceous of the Douala Basin, Cameroon, are examples of this kind of kerogen.

Intermediate types of organic matter may also occur, particularly those resulting from a mixture of Type II and Type III kerogens. This situation is found in marine environments where an autochthonous marine input derived from plankton is superimposed on a regional supply of terrestrial organic debris brought by rivers and currents. The resulting petroleum potential is intermediate.

The main types of kerogen are conveniently distinguished by using a van Krevelen-type diagram, where the atomic H/C and O/C ratios (resulting from elemental analysis) are plotted (Fig. 1). Each type of kerogen is located on the diagram along a specific line, called an *evolution path*. The increasing evolution of the organic matter, with increasing depth of burial and rising temperature, appears on the diagram as a change of composition along the same evolution path: the shallow, immature samples have original high H/C and O/C ratios; the deepest samples have low hydrogen and oxygen contents and they approach the origin of the diagram.

During this evolution, three successive stages may be distinguished, with increasing burial depth and temperature:

(a) *Diagenesis stage* is marked by a loss of oxygen, and consequently by a decrease of the O/C ratio, whereas the H/C ratio is slightly affected. In terms of petroleum exploration, diagenesis is an immature stage when mostly carbon dioxide, water, and some methane and heavy hetero-compounds are generated.

(b) *Catagenesis stage* is marked by a loss of hydrogen, and consequently by a decrease of the H/C ratio. The O/C ratio is usually not affected, or it may even increase, due to a preferential elimination of carbon and hydrogen. Catagenesis corresponds to the main stage of petroleum generation; first liquid oil, then wet gas and condensate.

(c) *Metagenesis stage* corresponds to deep or very deep burial. The H/C ratio is of the order of 0·5 and decreases only slightly. At that stage kerogen consists mostly of the residual polyaromatic nuclei, and only methane is generated, by cracking of the short alkyl chains.

Other Physico–Chemical Analyses
Other physico–chemical analyses carried out on kerogen may provide additional information on the type and the evolution stage of kerogen.

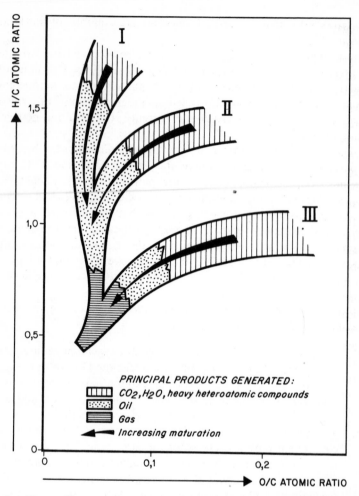

Fig. 1.   Types of kerogen according to their elemental composition, and principal products generated by thermal evolution. Reproduced from Ref. 28 by permission of Éditions Technip.

Some techniques like thermo-gravimetric analysis associated with mass spectrometry, electron micro-diffraction, and infrared spectrophotometry are powerful research tools and may be used to solve specific exploration problems. For instance the identification of the type of organic matter in advanced stages of maturation is usually difficult by chemical techniques; kerogen of any type has low H/C and O/C ratios, and contains mostly aggregates of polyaromatic nuclei. However, Oberlin et al.[15] have been able to show that the structure of these aggregates differs from one type of kerogen to another, by using *electron micro-diffraction*. In particular the size of the aggregates varies from less than 100 to 1000Å when passing from Type III to Type II, then Type I kerogen. *Infrared spectrophotometry*[19] may provide a comparative evaluation of the petroleum potential of different source rocks by recording the absorption bands related to aliphatic groups (source of hydrocarbons) and polyaromatic nuclei (inert part of kerogen).

Another physical technique used on kerogen is *electron spin resonance* (ESR), which has been proposed as a routine tool for measuring the degree of evolution of the source rocks, or even the palaeo-temperatures.[18] The parameter of the chemical structure of kerogens, which is responsible for the paramagnetic susceptibility and thus the ESR signal, is the abundance of free radicals. Free radicals appear in kerogen as a result of splitting bonds. For instance, elimination of an alkyl chain substituted on an aromatic ring results in the appearance of free radicals, until the two molecular fragments recombine (for instance with a hydrogen atom, or another alkyl chain). In general the abundance of free radicals increases with burial and temperature through to a certain stage of evolution, which approximately corresponds to a 2% vitrinite reflectance level.[9] At that stage the paramagnetic susceptibility is a maximum. With further burial and temperature increase, the signal decreases markedly (Fig. 2). In terms of kerogen structure it is likely that elimination of the alkyl groups substituted on polyaromatic nuclei (this elimination results in petroleum and gas generation) is mainly responsible for the signal increase. Beyond a maximum, which broadly corresponds to the end of wet gas and condensate generation, the signal decrease may be due to coalescence of the polyaromatic nuclei to form larger aggregates, resulting in the disappearance of free radicals.

Thus, a first drawback of ESR is the possibility for a given value of the signal to occur at two different stages of evolution (one above the maximum the other below it). A second problem is related to the influence of the kerogen type. Free radicals appearing on rigid polyaromatic nuclei are more stable than those appearing on other chemical groups. As a consequence, the same thermal history may result in a higher signal in Type III aromatic-rich kerogen and in a lower signal in Type II kerogen, and, moreover, in Type I aromatic-poor kerogen. This phenomenon has been effectively observed and could explain some apparent decrease of palaeo-temperature, interpreted from ESR, with increasing depth. In fact, facies changes may appear in the same manner as an increase or decrease of

B. TISSOT

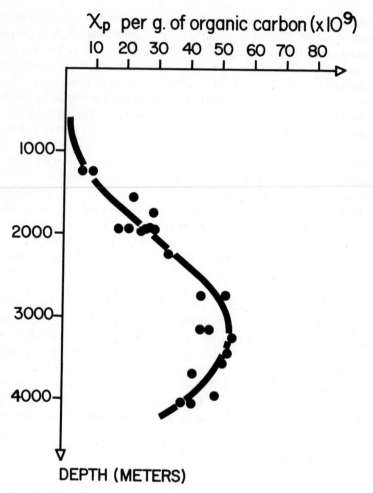

FIG. 2. Variation of paramagnetic susceptibility χ_p in Type III kerogen (Douala basin) as a function of depth. Reproduced from Ref. 9 by permission of Éditions Technip.

temperature. Finally, there may also be an age effect on the combination of free radicals. Hence, ESR may be a valid tool in situations where the type of organic matter remains constant, and where the level of 2% vitrinite reflectance is not reached. In other situations, an independent interpretation of ESR data seems difficult.

All methods based on kerogen analysis require the preliminary separation of kerogen from the mineral fraction..This step is rather lengthy and cannot always be achieved, due to a high residual content of pyrite in the kerogen. Furthermore, some of the analytical techniques are delicate or time consuming. For this reason, there is a need for a quick method using the sample of core or cuttings raw or after simple grinding. Pyrolysis assays seem to be a clue to the characterisation of the nature and degree of evolution of the organic matter. They can be used on a large number of samples for routine work and elaboration of geochemical well logs.

Pyrolysis Assays
Pyrolysis assays are carried out on small samples of cores or cuttings (about 100 mg) and require only grinding of the rock, without any further treatment. The most recently described method of assay[10] is outlined here.

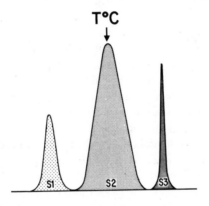

S2 ◗ *Hydrogen index*
S3 ◗ *Oxygen index*

FIG. 3.   Example of record obtained by pyrolysis of a source rock. Reproduced from Ref. 10 by permission of Éditions Technip.

The method uses a special device which heats the sample progressively, according to a pre-selected temperature programme, and measures four parameters during a 20-minute cycle (Fig. 3):

(a)   The hydrocarbons already present in the rock, which are volatilised by heating at moderate temperature (S_1);

(b) The hydrocarbon-type compounds generated during pyrolysis of the kerogen present in the rock (S_2);

(c) The oxygen-containing volatiles generated during pyrolysis: carbon dioxide and water. The commercial version uses only CO_2. It is measured in a convenient temperature window, in order to fulfil two requirements: to cover the main stage of CO_2 generation from organic matter and to avoid any other source of CO_2, such as the degradation of carbonates. In that respect, the upper limit of the window is imposed by siderite, which is the most labile common carbonate on pyrolysis. This parameter is called S_3;

(d) The temperature corresponding to the peak of hydrocarbon generation by pyrolysis (T).

From these measurements, it is possible to compute five parameters which, in turn, allow the determination of the type and the evolution stage of the organic matter.

Type and petroleum potential of the source rocks: When related to the total organic content (org. C) the values S_1 and S_2 become independent of the abundance of organic matter. The ratios S_1/org. C and S_2/org. C are called the *hydrogen index* and *oxygen index*, respectively. Analyses performed on various types of rock have demonstrated the existence of a fairly good correlation on the one hand between the hydrogen index and the H/C ratio of the kerogen, and on the other hand, between the oxygen index and the O/C ratio of the kerogen. Thus, the hydrogen index and oxygen index can be used to produce a van Krevelen-type plot (Fig. 4) and to differentiate the main types of organic matter. Furthermore, a quantitative evaluation of the *genetic potential* can be made. The quantities S_1 and S_2 represent the respective amounts of petroleum that have already been generated (S_1) and would be generated (S_2) upon further burial and temperature increase. Thus $S_1 + S_2$ is an evaluation of the genetic potential, expressed in kilograms of petroleum (oil plus gas) per tonne of rock.

Evolution stage of the source rocks: The temperature T (corresponding to the maximum of the pyrolysis peak) depends on the type of organic matter in shallow immature samples. With increasing depth and temperature, T increases and becomes progressively the same for all types of kerogen (Fig. 5). Thus, T can be used as a scale of source rock evolution. It may be corrected with respect to the type of organic matter in the low evolution stages.

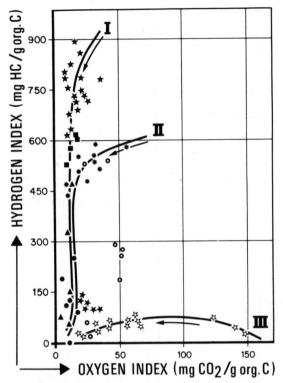

★   *Green River shales*

∗   *Lower Toarcian, Paris Basin*

▲   *Silurian-Devonian, Sahara-Libya*

●   *Upper Paleozoic, Spitsbergen*

☆   *Upper Cretaceous, Douala Basin*

■   *Cretaceous, Persian Gulf (Oligostegines limestone)*

○   *Upper Jurassic, North Aquitaine*

FIG. 4.   Classification of source rocks, according to pyrolysis indices. The Types I, II and III correspond to the three main types of kerogen on Fig. 1. Reproduced from Ref. 10 by permission of Éditions Technip.

A quantitative evaluation of the *transformation ratio* may be achieved by using the ratio $S_1/(S_1 + S_2)$ which is the ratio of the hydrocarbons already generated to the total genetic potential. The two parameters, T and $S_1/(S_1 + S_2)$ had been proposed previously by Barker.[3] They can be calibrated against any other classic evolution index, e.g. vitrinite reflectance.

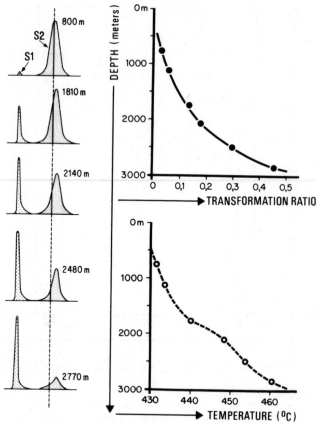

FIG. 5. Characterisation of the maturation of source rocks by pyrolysis in sediments of Tertiary age. *Top:* transformation ratio $S_1/(S_1 + S_2)$. *Bottom:* temperature T corresponding to the peak of hydrocarbon generation during pyrolysis. Reproduced from Ref. 10 by permission of Éditions Technip.

The pyrolysis assay described here, according to Espitalié *et al.*,[10] takes advantage of fundamental studies based on kerogen elemental composition, owing to the correlation of the hydrogen and oxygen indices with the H/C and O/C ratios. However, due to the absence of preparative treatment (except grinding) and to the rapidity of the analysis (one sample every twenty minutes) it is possible to run a large number of samples and to establish geochemical logs of the wells drilled. Pyrolysis could even be performed on the well site and be a step towards on-time geochemical logging.

Chemical Analyses of the Bitumen

Chemical analyses of the bitumen (hydrocarbons plus resins and asphaltenes) obtained by solvent extraction of source rocks have also been used as a tool to characterise the evolution stage of the organic matter. The bitumen/total organic carbon ratio and the hydrocarbon/total organic carbon ratio are directly linked to the transformation ratio of the organic matter. However, the ratios based on solvent extraction do not account for light hydrocarbons which are lost during the analytical procedure. Furthermore, migration phenomena may affect them. Nevertheless, if a sufficient number of rock samples from the same formation are available at different depths, the hydrocarbon/total organic carbon ratio can be plotted against depth and the general curve of petroleum generation may be traced (Fig. 6). Thus, the oil window is defined, overlain by the immature zone and underlain by the gas zone.

Unfortunately, it is often difficult to obtain samples conveniently spread over a large depth range. For these reasons, several indices based on hydrocarbon composition have been proposed. Broadly speaking, they are all based on the progressive change that leads from biogenic hydrocarbons with characteristic distributions at shallow depths, to more simple, low molecular weight hydrocarbons generated by thermal degradation of kerogen at greater depths. The latter distributions are comparable to the distribution observed in crude oil. For example, a *Carbon Preference Index* (CPI) has been proposed,[4] based on the ratio of odd-numbered n-alkanes to even-numbered n-alkanes. At shallow depth, the odd-numbered alkanes are fairly predominant, because this condition is found in the cuticular waxes of terrestrial plants and also in algal lipids. Thermal degradation of kerogen during catagenesis results in a new alkane generation without odd or even preference. Thus, the original molecules are progressively diluted by the more abundant new hydrocarbons and the CPI decreases from values as high as 5 in recent sediments to about 1 in the main zone of oil formation. Other indices like the *naphthene index*[17] are based on comparable considerations. Polycyclic naphthenes derived from sterols and triterpenes account for a large proportion of the saturated hydrocarbons at shallow depths. They are progressively diluted by one- or two-ring naphthenes generated from kerogen at greater depths.

Lighter fractions of the hydrocarbons disseminated in source rocks have also been used. With increasing maturation, the content of normal and branched alkanes usually increases relative to naphthenes and aromatics. Philippi[17] used this technique in the C_7 range; the C_7 hydrocarbons are separated into alkanes, naphthenes and aromatics, and plotted on a

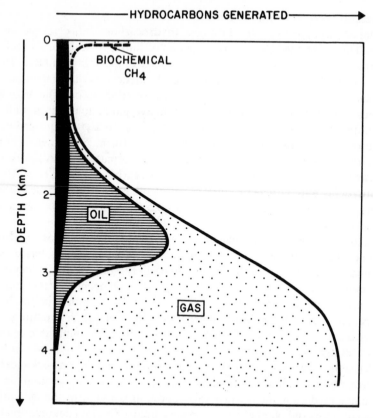

FIG. 6. General scheme of hydrocarbon generation as a function of burial depth. The depth scale depends on the type of kerogen and the geothermal gradient. Reproduced from Ref. 23 by permission of Association Française des Techniciens du Petrole.

triangular diagram. Increasing evolution is denoted by a move towards the alkane pole of the diagram. In fact, this property is not restricted to the C_6 or C_7 hydrocarbons, and a comparable plot (alkanes, naphthenes, aromatics) of the total C_{15+} fraction extracted by a solvent provides the same information. Jonathan et al.[14] used a somewhat different technique, based on the same considerations. By thermal volatilisation, the light hydrocarbons are expelled from the rock and trapped, then released and gas chromatographed. The ratios of some peaks conveniently selected— like n-hexane to methylcyclopentane—are used as maturation indices. The ratios can be calibrated against other parameters, such as vitrinite reflectance. Finally, Bailey *et al.*[2] remark that the light hydrocarbons C_4–C_7 are absent above the oil window and again at depth, below the wet gas zone. Thus, they plot the total content of those light hydrocarbons versus depth.

Optical Examination of Organic Matter

Optical examination of the organic matter provides valuable information on the type and the evolution stage of the kerogen. In addition, it is possible to recognise mixing of different organic sources, either contemporaneous (e.g. marine phytoplankton and terrestrial plants) or reworked from previous sedimentary cycles.

Microscopic examination of kerogen preparations in *transmitted light* is derived from palynological examinations.[5] Classification of kerogen type is based mostly on the general shape of the organic remnants, because formally identifiable micro-fossils account for only a minor part of the organic matter. Two main types are usually distinguished: 'humic' material, supposedly of herbaceous or woody origin, where shape or structure of the vegetal tissues is still recognisable—it is usually considered to be a gas source rock and a poor oil source rock; and 'amorphous' material with a vaporous or cloudy shape, and no identifiable structure, is commonly referred to as sapropelic organic matter and considered to be a good source rock.

When comparing this classification with the types of kerogen based on chemical composition, there seems to be a broad agreement, although certain formations are classified differently. Humic material usually belongs to kerogen Type III. Most of the material considered as sapropelic refers to kerogen Type II, but some has a chemical composition comparable to Type III.

Besides allowing classification of the organic matter, examination of kerogen in transmitted light also provides a scale of evolution based on colour and state of preservation of spore and pollen material. Originally these organic remnants are yellow. With increasing evolution they become successively orange, brown and finally black. Based on these considerations, semi-quantitative scales of evolution have been proposed, ranging from 1 (yellow) to 5 (black).[6,11,20]

Microscopic examination of the reflectance of polished sections is derived from coal petrology. In the same way as coal petrographers distinguish coal macerals, the organic constituents are grouped in three main classes of increasing reflectance: liptinite, vitrinite and inertinite. The varying proportions of these constituents are an indication of the origin and composition of the organic matter. However, the determinations are mostly concerned with the larger fragments of organic matter and those are not necessarily representative of the total organic fraction.

Reflectance measurements performed on vitrinite have long been a way to measure the evolution of coals. It has been subsequently adapted to disseminated organic matter and widely used for this purpose.[21,13] In the diagenesis stage, the reflectance of vitrinite—or its precursors like huminite—is low, i.e. below 0·5%, and changes slowly with increasing burial. During catagenesis, reflectance increases more rapidly as a function of depth, and rises from 0·5 to about 2%. Later, in the metagenesis stage, the reflectance increase continues. Metamorphism, as defined by the occurrence of greenschists, is reached for values as high as 4% (Fig. 7).

Vitrinite reflectance is certainly the most widely used optical technique for characterising kerogen evolution, and it is possibly the best facility offered in respect of microscopic examination. There are, however, several difficulties which should be known and taken care of. First, some preparations show different types of 'vitrinite', with different reflectances. Among them, the population with the lowest reflectance is usually considered as autochthonous, whereas the other types are considered as reworked. However, there are circumstances where such a reworking is difficult to conciliate with geological considerations. On the other hand, vitrinite fragments are easy to find in Type III kerogen, but they may be rare or absent in certain kerogens of Type I or II.

Finally, there has been some controversy about the boundaries between the successive immature, oil, wet gas and condensate, and dry gas zones, in terms of the reflectance scale. In fact, thermal degradation of kerogen, to produce hydrocarbons, always obeys the kinetics laws. However, the parameters of these laws, such as activation energies, have different distributions in different kerogens. In other words, the same thermal history—and thus the same vitrinite reflectance—may result in a different situation with respect to oil generation, according to the type of organic matter. The differences are likely to be slight with respect to the definition of the gas zone. Broadly speaking, the beginning of the main zone of oil generation may start in the range of 0·5% to 0·7% (earlier in Type III kerogen, later in Type I kerogen), and terminate at 1·3% or somewhat earlier, according to the kerogen type. Then, wet gas and condensate may extend up to about 2%; beyond that limit is the dry gas zone.[29]

Fluorescence due to excitation by UV light is a way of locating the organic matter on sections. It may also be used to characterise the evolution of kerogen; fluorescence is high in shallow immature samples and decreases

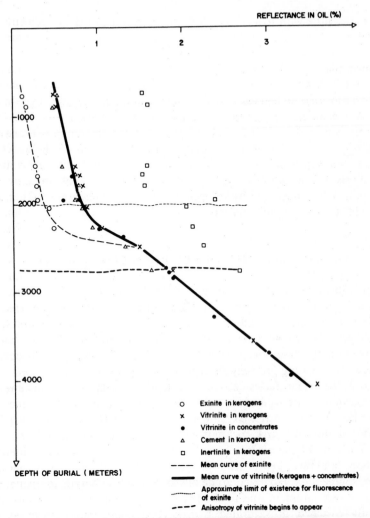

REFLECTANCE IN OIL (%)

DEPTH OF BURIAL (METERS)

o	Exinite in kerogens
x	Vitrinite in kerogens
•	Vitrinite in concentrates
△	Cement in kerogens
□	Inertinite in kerogens
–––	Mean curve of exinite
▬▬▬	Mean curve of vitrinite (kerogens + concentrates)
....	Approximate limit of existence for fluorescence of exinite
‒ ‒ ‒	Anisotropy of vitrinite begins to appear

FIG. 7. Reflectance of the main constituents of Logbaba organic matter as a function of the depth of burial. Reproduced from Ref. 8 by permission of Pergamon Press.

in the main zone of oil generation.[1] It drops to zero at the end of the wet gas zone. Furthermore, the wavelength range of the fluorescence light progressively moves towards the red with increasing burial.[16] Fluorescence is a promising field for the characterisation of kerogen evolution, since its range of variation is mostly located over the zone where vitrinite reflectance increases little with increasing depth (i.e. beginning of the oil generation zone).

Conclusion

As a conclusion on source rock evaluation, it can be said that the association of one chemical and one optical technique is always advisable in order to characterise the type of organic matter and the stage of evolution. Pyrolysis provides a fast evaluation of both parameters, and is probably one of the best. This operation may even be performed on the well site and it is a step towards on-time geochemical logging. In the laboratory, pyrolysis can be complemented by optical examination (palynofacies and vitrinite reflectance), and by kerogen preparation and analysis for a more refined study of organic matter.

OIL/OIL AND OIL/SOURCE ROCK CORRELATION

In petroleum exploration, identification of the various types of oil or gas occurring in a sedimentary basin is of great interest. It helps the geologist to define the number of different targets, which are present in different geological horizons. This result is achieved by oil/oil or gas/gas correlation. Furthermore, the oil/source rock relationship is of special interest for locating precisely the formation responsible for oil generation. In turn, the extent, facies changes and depth of burial of the source rock formation may be studied across the basin. The problems of correlation have been recently reviewed by Welte et al.[31] and Deroo.[7]

Oil/oil and oil/source rock correlations utilise the same methods, i.e. comparisons of the chemical composition of various classes of hydrocarbons and non-hydrocarbons in crude oils and bitumen extracted from source rock. However, crude oils are severely changed, compared to source rock bitumen, due to the migration phenomena. Crude oils are mostly enriched in saturated hydrocarbons, moderately enriched in aromatics and depleted in N, S and O compounds, compared to source rock bitumen (Fig. 8). Furthermore, low molecular weight hydrocarbons are also favoured over the heavier molecules. Thus bulk compositional parameters—e.g. saturates, aromatics and N, S, and O content—might be used for oil/oil correlation, but are useless for oil/source rock correlation.

The best tools for correlation are provided by geochemical fossils, i.e. molecules synthesised by organisms at the time of sediment deposition and preserved unchanged, or with minor alteration, during geological history.

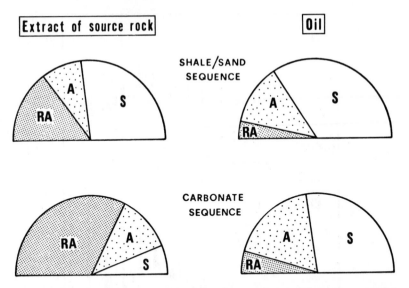

FIG. 8.   Change of petroleum composition due to migration from source rock to reservoir. RA = Resins and Asphaltenes, A = Aromatic HC, and S = Saturated HC.

These molecules usually have a sufficiently characteristic distribution to permit differentiation between individual source beds and crude oils. However, owing to the possible effects of migration, it is advisable to make use of the distribution within a group of molecules having comparable physical properties (polycyclic naphthenes or *n*-alkanes, or isoprenoids) and not to compare the amounts of very different molecules (e.g. *n*-alkanes/polycyclic naphthenes).

The most commonly used geochemical fossils are *n*-alkanes, isoalkanes including isoprenoids, polycyclic naphthenes, polycyclic naphtheno-aromatics and aromatics. This order broadly corresponds to increasing complexity of chemical make-up. In addition, some non-biogenic compounds like benzothiophene derivatives may be present in both oil and source rock and thus be a valuable tool for high-sulphur crude oils.

Gas chromatography (GC) of the total saturated fraction provides a finger-print of the bitumen or crude oil and allows a calculation of the distribution of *n*-alkanes and isoprenoid isoalkanes. More refinement is obtained by preliminary separation of the linear and branched + cyclic fractions, which are separately chromatographed. Other isoalkanes and polycyclic naphthenes (steranes and triterpanes) distributions can

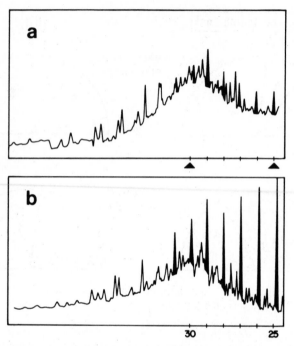

FIG. 9.　Oil/source rock correlation based on chromatographic finger-print of high molecular weight saturated hydrocarbons (Aquitaine Basin). (a) Crude oil, Le Puch, Lower Aptian, and (b) Rock extract, Mano, Upper Jurassic. Reproduced from Ref. 7 by permission of Société Nationale Elf Aquitaine.

only be obtained after preliminary separation by using the combination of gas chromatography and mass spectrometry (GC–MS).

An example of correlation based on the finger-printing of total saturated hydrocarbons is shown in Fig. 9, and is self-explanatory. An example of correlation based on a detailed study of four- and five-ring naphthenes comprising 27 to 30 carbon atoms is shown in Fig. 10. The correlation is good for the oil/rock pairs from the Uinta Basin, Utah, and the Midland Basin, Texas. In the Rhine Graben area, Germany, the oil/source rock correlation is fair for the Oligocene oil; it is also fair between the Eocene oil and the Middle Jurassic source rock K-1, but the same oil shows no correlation with the Lower Jurassic sources K-2 or K-3. Evaluating the information provided for correlation by analysis of saturated hydrocarbons, Welte et al.[31] placed more emphasis on polycyclic naphthenes, which exhibit specific molecular structure, and less on n-alkanes or

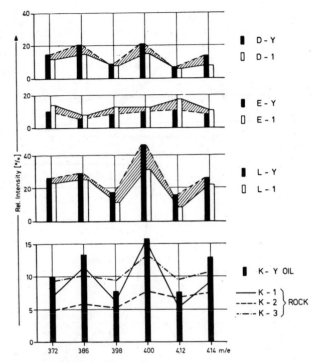

FIG. 10. Comparison between correlative source rocks and oils with respect to the distribution of averaged molecular ion series of C_{27+} cyclics.[31]
D-Y (oil) D-1 (rock): Oligocene, Rhine Graben, West Germany.
E-Y (oil) E-1 (rock): Eocene, Uinta Basin, Utah, USA.
L-Y (oil) L-1 (rock): Permian, Midland, Texas, USA.
K-Y (oil): Eocene
K-1 (rock): Middle Jurassic   } Rhine Graben, West Germany.
K-2 and K-3 (rocks): Lower Jurassic

isoprenoids, which are more dependent on temperature history. In fact, a smooth distribution of n-alkanes, regularly decreasing with increasing carbon number is a very common feature found in many crude oils around the world.

Saturated hydrocarbons become less and less effective for correlation with increasing maturation (age and depth). Distributions of n-alkanes become very similar, and polycyclic naphthenes become too scarce for separation and identification. The Palaeozoic crude oils from the Illizi Basin (Algeria) are an example of this situation. Global analysis of crude oil, gas chromatography and mass spectrometry of the saturated

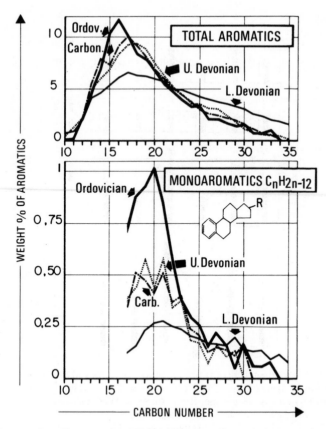

FIG. 11. Crude oil types in Illizi basin, based on analysis of aromatic hydrocarbons by mass spectrometry. *Top:* distribution of total aromatic fraction by carbon number. *Bottom:* the same distribution for monoaromatic steroids. Reproduced from Ref. 26 by permission of Éditions Technip.

hydrocarbons do not allow any distinction to be made between oils reservoired in Ordovician, Lower Devonian, Upper Devonian and Carboniferous reservoirs. However, a detailed study by mass spectrometry of the aromatic and naphtheno-aromatic hydrocarbons provides a fair differentiation between three different groups of oil (Fig. 11). This result is obtained mainly by consideration of the four-ring molecules derived from steroids. Although the saturated four-ring molecules are practically no longer detectable, aromatised molecules comprising one aromatic and three naphthenic rings are more stable and thus they have been preserved even in old crude oils and source rocks. The distribution of these molecules, obtained by mass spectrometry, made the oil/source rock correlation possible.

Carbon isotope distribution has been used as a correlation tool for oil/oil and gas/gas relationships, and also for identification of the related source rock. The isotope ratio is directly used for comparing oils or gases. With respect to oil/source rock correlation, it has been observed in many cases that the carbon of the source rock bitumen is isotopically lighter than the related kerogen. In turn, most crude oils are isotopically lighter, or very similar to, the source rock bitumen.[31] Based on these considerations, the difference in isotopic composition between an oil and the kerogen or the bitumen of a supposed source rock may be of diagnostic value.

TIMING OF HYDROCARBON GENERATION

The introduction of geological time into our reasoning is of prime importance for petroleum exploration. Knowledge of the time of oil or gas generation allows one to compare it with the age of trap development; for example, the age of the anticlinal folds, of the faults, and of the deposition of an impermeable cover over an unconformity surface.

Geological time seldom has been explicitly taken into account. In particular the role of depth of burial and temperature as a function of time has not been considered except in a small number of cases. However, the reactions of hydrocarbon generation obey the laws of kinetics, in which the quantity of products formed depends not only on temperature, but also on the reaction time. Intuitively, it is clear that the quantity of hydrocarbons generated from a given kerogen will not be the same after one million years as after a hundred million years, if the kerogen is heated to the same temperature.

There are two approaches to the timing of hydrocarbon generation. One is semi-empirical and based on the comparison of burial and oil generation graphs. It is mostly valid in areas with a simple burial/time relationship. The other uses a mathematical model of hydrocarbon generation. The model allows for complex burial/time relationships (e.g. two cycles of subsidence separated by folding and erosion), for variations of geothermal gradient, etc. Thus, it is of a more general use, and it makes possible a quantitative evaluation of oil and gas generated as a function of time.

The first approach is based on the hydrocarbon generation curve (Fig. 6). If a sufficient number of samples of the source rock can be obtained from

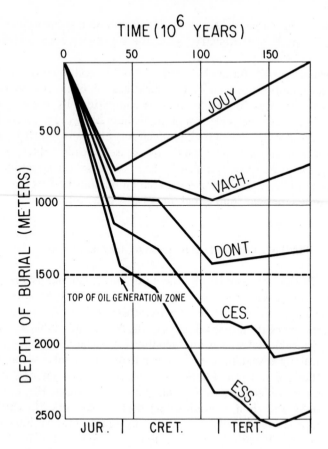

FIG. 12.   History of burial and time of oil generation in the lower Toarcian shales of the Paris basin. Sample localities: JOUY = Jouy; VACH. = Vacherauville; DONT. = Dontrian; CES. = Césarville; ESS. = Essises.

various depths, the oil generation curve can be plotted as a function of depth. Thus, it becomes possible to mark two depth thresholds: the top of the main zone of oil generation, and its bottom, i.e. the top of the gas generation zone. It is then possible to transfer the depth thresholds on to the burial graph (plot of burial depth versus time for each location in the basin). The times at which the burial curve crosses these depths are the times of the beginnings of important oil generation and of important gas generation, respectively. For example, in the Paris Basin (Fig. 12) important oil generation from Toarcian shales began in Lower Cretaceous time at Essises, and in Late Cretaceous only at Césarville.

The second approach is based on a mathematical model of oil and gas generation described by Tissot[22] and Tissot and Espitalié.[28] The simulation of oil and gas formation takes into account the nature of the kerogen, the history of burial, and the geothermal gradient. The nature of kerogen is represented by the distribution of the activation energies used in the model. They are adjusted on the basis of observations (petroleum compounds extracted with solvents from well samples) and data obtained from laboratory tests (pyrolysis). They may also be adjusted on the basis of pyrolysis data alone. The history of burial is reconstructed by means of isobath and isopach maps based on well logs. By using a regular grid, a curve is drawn giving the depth of the source rock as a function of time, for each point of the network. The problem of reconstructing the eroded thickness is resolved, in general, by extrapolation from the deepest zones of the basin.

Determining the geothermal gradient is more difficult. In general high gradients are observed in the areas of rifting, and in the zones where folding and magmatic events are young. In areas where these phenomena are old, the gradients seem to be stabilised. For example, one can reasonably use the present-day gradients for the Jurassic, Cretaceous and Tertiary in provinces where the last orogeny was in the Palaeozoic or earlier. At other locations (e.g. basins on continental margins bordering the Atlantic Ocean) the ancient geothermal gradient can be calculated. A parameter sensitive to thermal evolution, e.g. vitrinite reflectance, is measured on a certain number of samples from a well. Simulation is then used to calculate the same values of reflectance, and the geothermal gradient is adjusted by successive iterations until the difference between the calculated and the measured values is minimised. The method is described by Tissot and Espitalié.[28]

The main reason for using the mathematical model, rather than the graphic plot, is to handle more complex situations, e.g. changing geothermal gradient or successive cycles of sedimentation separated by folding and erosion. An example of this last situation is the Hassi-Messaoud area in Algeria. The northern Sahara Basin shows two cycles of sedimentation. The Palaeozoic system contains sandstones and shales of Cambrian to Carboniferous age, among which the Lower Silurian shales are the source rock and the Cambrian and Ordovician sandstones are the reservoirs of the oils. Moderate Hercynian folding was followed by major erosion which left only Cambrian or Ordovician reservoirs on the anticlines. Sedimentation was resumed in Triassic time and resulted in important Mesozoic deposits, including a thick Triassic–Jurassic salt, which actually seals the Cambro–Ordovician oil reservoirs (Fig. 13). This situation poses the problem of the timing of oil accumulation. If the petroleum was generated and accumulated in the Hassi-Messaoud dome

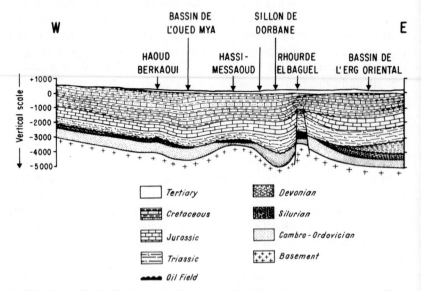

FIG. 13.   Geological cross-section in the Hassi-Messaoud area, Algeria.[27]

during the Palaeozoic, it was exposed to the atmosphere because of erosion for millions of years at the end of the Carboniferous and during the Permian. Oil should have been either destroyed or converted to heavy oils and tars. If, on the contrary, the accumulation was formed after the Triassic, the oil could have been trapped and protected by the salt beds, which constitute an effective cover.

Figure 14 shows the burial of the Lower Silurian source rocks, as a function of depth, in one of the synclines bordering the Hassi-Messaoud dome. Below, the formation of oil and gas is shown, as calculated by the mathematical model. Little oil was generated in the course of the first 300 million years, and the principal stage of oil formation was not reached until the Cretaceous. Thus, accumulation of oil in the reservoirs occurred after the domes had been sealed by the salt cover.

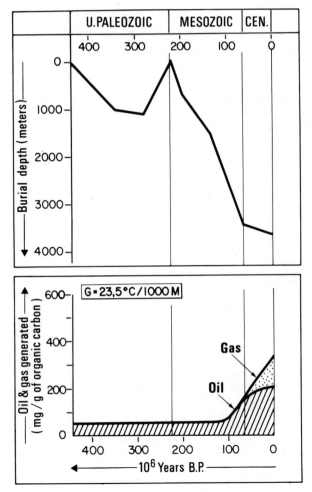

FIG. 14.   Reconstitution of burial depth (*top*) and mathematical simulation of oil generation (*bottom*) in the Hassi-Messaoud area.[27]

CONCLUSION

Advances in organic geochemistry during the last ten years have resulted in understanding the importance of the type of organic matter and of the thermal evolution during burial. It is now possible to identify the different potential source rocks occurring in a sedimentary basin, and to quantify their respective potentials by laboratory tests. The stage of maturation is currently evaluated by various physico–chemical and optical techniques. Furthermore, the oil/source rock relationship can be established in most cases, based on detailed analysis of crude oil and source rock hydrocarbons.

By the association of these various techniques, organic geochemistry is now in a position to offer a systematic appraisal of the petroleum potential of a sedimentary basin. Introduction of fast but powerful methods like pyrolysis could make it possible, in the near future, to transfer many geochemical analysis techniques from the central laboratory to the well site. Some kind of on-time geochemical logging may even be forecast, allowing early decisions on the future of the exploration programme.

REFERENCES

1. ALPERN, B., DURAND, B., ESPITALIÉ J., and TISSOT, B. (1972). Localisation, caractérisation et classification pétrographique des substances organiques sédimentaires fossiles, in: *Advances in Organic Geochemistry* 1971, Pergamon Press, Oxford-Braunschweig (1972) 1–28.
2. BAILEY, N. J. L., EVANS, C. R. and MILNER, C. W. D. (1974). Applying petroleum geochemistry to search for oil: examples from Western Canada Basin, *Am. Assoc. Petrol. Geologists Bull.*, **58**, 2284–94.
3. BARKER, C. (1974). Pyrolysis techniques for source-rock evaluation, *Am. Assoc. Petrol. Geologists Bull.*, **58**, 2349–61.
4. BRAY, E. E. and EVANS, E. D. (1961). Distribution of *n*-paraffins as a clue to recognition of source beds, *Geochim. Cosmochim. Acta*, **22**, 2–15.
5. COMBAZ, A. (1964). Les palynofacies, *Rev. Micropaleontologie*, 3.
6. CORREIA, M. (1967). Relations possibles entre l'état de conservation des éléments figurés de la matière organique (microfossiles palynoplanctologiques) et l'existence de gisements d'hydrocarbures, *Inst. Français Pétrole Rev.*, **22**, 1285–1306.
7. DEROO, G. (1976). Corrélations huiles brutes—roches mères à l'échelle des bassins sédimentaires. *Bull. Centre Rech. Pau*, **10**, 317–35.
8. DURAND, B. and ESPITALIÉ, J. (1976). Geochemical studies on the organic matter from the Douala Basin (Cameroon) II. Evolution of kerogen, *Geochim. Cosmochim. Acta*, **40**, 801–8.
9. DURAND, B., MARCHAND, A. and COMBAZ, A. (1977). Etude de kérogènes en résonance paramagnétique électronique in: *Proceedings of the 7th International Congress of Organic Geochemistry* 1975 (J. Goni and R. Campos, eds.), Technip, Paris (in press).
10. ESPITALIÉ J., LAPORTE, J. L., MADEC, M., MARQUIS, F., LEPLAT, P., PAULET, J. and BOUTEFEU, A. (1977). Méthode rapide de caractérisation des roches mères, de leur potentiel pétrolier et de leur degré d'évolution, *Inst. Français Pétrole Rev.*, **32**, 23–42.

11. GUTJAHR, C. C. M. (1966). Carbonization of pollen grains and spores and their application, *Leidse Geologische Mededelingen*, **38**, 1–30.
12. HOOD, A. and CASTANO, J. R. (1974). Organic metamorphism. Its relationship to petroleum generation and application to studies of authigenic minerals, *Coordinating Comm. Offshore Prospecting Techn. Bull.*, **8**, 85–118.
13. HOOD, A., GUTJAHR, C. C. M. and HEACOCK, R. L. (1975). Organic metamorphism and the generation of petroleum, *Am. Assoc. Petrol. Geologists Bull.*, **59**, 986–96.
14. JONATHAN, D., L'HOTE, G. and DU ROUCHET, J. (1975). Analyse géochimique des hydrocarbures légers par thermovaporisation, *Inst. Français Pétrole Rev.*, **30**, 65–88.
15. OBERLIN, A., BOULMIER, J. L. and DURAND, B. (1974). Electron microscope investigation of the structure of naturally and artificially metamorphosed kerogen, *Geochim. Cosmochim. Acta*, **38**, 647–9.
16. OTTENJANN, K., TEICHMULLER, M. and WOLF, M. (1974). Spektrale Fluoreszenz-Messungen an Sporiniten mit Auflicht-Anregung, eine mikroskopische Methode zur Bestimmung des Inkohlungsgrades gering inkohlter Kohlen, *Fortschr. Geol. Rheinl. u Westf.*, **24**, 1–36.
17. PHILIPPI, G. T. (1965). On the depth, time and mechanism of petroleum generation, *Geochim. Cosmochim. Acta*, **29**, 1021–49.
18. PUSEY, W. C. (1973). How to evaluate potential gas and oil source rock. *World Oil*, **176**(5), 71–5.
19. ROBIN, P. L., ROUXHET, P. G. and DURAND, B. (1977). Caractérisation des kérogènes et de leur évolution par spectroscopie infra-rouge, in: *Proceedings of the 7th International Congress of Organic Geochemistry* 1975 (J. Goni and R. Campos, eds.) (in press).
20. STAPLIN, F. L. (1969). Sedimentary organic matter, organic metamorphism, and oil and gas occurrence. *Canadian Petrol. Geologists, Bull.*, **17**, 47–66.
21. TEICHMULLER, M. (1971). Anwendung kohlenpetrographischer methoden bei der erdol und erdgasprospektion, *Erdöl und Kohle*, **24**, 69–76.
22. TISSOT, B. (1969). Premières données sur les mécanismes et la cinétique de la formation du pétrole dans les sédiments. Simulation d'un schéma réactionnel sur ordinateur, *Inst. Français Pétrole Rev.*, **24**(4), 470–501.
23. TISSOT, B. (1973). Vers l'évaluation quantitative du pétrole formé dans les bassins sédimentaires, *Revue Assoc. Fr. Tech. Pétrole*, **222**, 27–31.
24. TISSOT, B. and PELET, R. (1971). Nouvelles données sur les mécanismes de genèse et de migration du pétrole. Simulation mathématique et application à la prospection, *Proc. 8th World Petroleum Congress*, **2**, 35–46.
25. TISSOT, B., DURAND, B., ESPITALIÉ, J. and COMBAZ, A. (1974). Influence of nature of diagenesis of organic matter in formation of petroleum, *Am. Assoc. Petrol. Geologists Bull.*, **58**, 499–506
26. TISSOT, B., ESPITALIÉ, J., DEROO, G., TEMPERE, C. and JONATHAN, D. (1974). Origine et migration des hydrocarbures dans le Sahara oriental (Algérie), in: *Advances in Organic Geochemistry* 1973 (B. Tissot and F. Bienner, eds), Technip, Paris, 315–34.
27. TISSOT, B., DEROO, G. and ESPITALIÉ, J. (1975). Etude comparée de l'époque de formation et d'expulsion du pétrole dans diverses provinces géologiques, *Proc. 9th World Petroleum Congress*, **2**, 159–69.

28. TISSOT, B. and ESPITALIÉ, J. (1975). L'évolution thermique de la matière organique des sédiments: application d'une simulation mathématique, *Inst. Français Pétrole Rev.*, **30**, 743–77.
29. VASSOEVICH, N. B., AKRAMKHODZHAEV, A. M. and GEODEKYAN, A. A. (1974). Principal zone of oil formation, in: *Advances in Organic Geochemistry* 1973 (B. Tissot and F. Bienner, eds.) Technip, Paris, 309–14.
30. WELTE, D. H. (1972). Petroleum exploration and organic geochemistry, *J. Geochem. Explor.*, **1**, 117–36.
31. WELTE, D. H., HAGEMANN, H. W., HOLLERBACH, A., LEYTHAEUSER, D. and STAHL, W. (1975). Correlation between petroleum and source rock, *Proc. 9th World Petroleum Congress*, **2**, 179–91.

Chapter 3

PETROLEUM MIGRATION AND ACCUMULATION

KINJI MAGARA

Imperial Oil Ltd, Alberta, Canada

SUMMARY

Four possible causes of primary petroleum migration are discussed in this chapter: (a) forces resulting from sediment compaction (movement of compaction water); (b) the water expansion effect due to temperature increase during burial; (c) the osmotic effect; and (d) montmorillonite–illite conversion (montmorillonite dehydration). The relative importance of these causes may be related to another problem—the form of oil or gas at the time of primary migration; whether it is in solution in water or in its own phase.

The author judges that a large proportion of migrating oil must be in its own phase because of the relatively low solubility. But most of the migrating gas must be in solution in water.

The final part of this chapter discusses the importance of pressure and capillary seals in the accumulation of oil and gas in the traps.

INTRODUCTION

Most petroleum hydrocarbons have been generated by a subsurface thermal process from the organic matter deposited along with fine-grained clastic sediments. The thermal process can be expressed by a geological time–temperature function. Connan[6] derived such a function by examining accumulated oils of known geological ages and temperatures, and a possible oil-generation zone in terms of geological time and temperature was proposed by Hunt.[18]

Hydrocarbons generated in fine-grained sedimentary rocks are probably disseminated at first, but eventually they must move from their host rocks into more permeable and porous sediments to form an accumulation.

The movement of petroleum from non-reservoir rocks to reservoir rocks is called *primary migration*, and is distinguished from its concentration and accumulation within the reservoir rocks into pools of oil and gas, known as *secondary migration*.[21]

PRIMARY MIGRATION

A discussion of the primary migration of hydrocarbons is concerned with three different kinds of problems:

1.   The form in which they migrate, such as molecular solution, micellar solution and a separate hydrocarbon phase.
2.   The cause of migration.
3.   The water source and the cause of its movement.

Whether the hydrocarbons being discussed are mainly gas or mainly liquid will affect the proportions of hydrocarbons that are moving in solution and as a separate phase, because the solubility of gaseous hydrocarbons is generally higher. The migration mechanism for hydrocarbons as a separate phase may differ from that for hydrocarbons in solution in water. If water movement is important in moving hydrocarbons, the source of the water and the cause of its movement must be examined carefully.

The molecular solubility of liquid hydrocarbons in water at relatively high temperatures was recently discussed by Price,[36] who showed that the solubility increased with increasing temperature (Fig. 1). The solubility of the Farmers' oil at 160 °C is approximately 150 ppm, and the curve shows that it tends to increase with further increase in temperature. However, these temperature values are much higher than the known temperature range of 60° to 150 °C for active oil generation. Dickey,[7] on the other hand, suggested that the flowing stream would have to contain at least 10 000 ppm of hydrocarbons at the time of primary migration. Therefore, it may be very difficult to believe that most oil migrates as a molecular solution in water.

In the case of the migration of gas, the situation can be completely different. According to Dodson and Standing,[9] the solubility of natural gas in water ranges from 4 cubic feet per barrel of water at 400 psi, up to 22

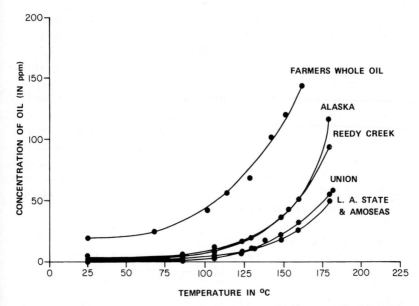

FIG. 1.   Solubilities of two whole oils (Wyoming Farmers and Louisiana State) and four topped oils (Amoseas Lake, Reedy Creek, Alaska and Union Moonie) as functions of temperature in water. Topping temperature is 200 °C (392 °F). Reproduced from Ref. 36 by permission of the American Association of Petroleum Geologists.

cubic feet at from 2000 to 6000 psi. In other words, most of the gas might migrate in solution in water at the primary migration stage.

Another solution mechanism, called micellar solution, was proposed by Baker.[2] He suggested that hydrocarbon solubility is substantial if the water contains micelles formed by the soaps of organic acids. However, there are several reasons why Baker's proposal is not plausible as the main mechanism of hydrocarbon migration in the subsurface. First, there is no good evidence that such solubilising micelles exist in significant quantities in shales. Even if they do exist in shales, they would not be easily moved because they are not small. Moreover, the micelles would increase the solubility of the heavier hydrocarbons in water only to a few parts per million—nowhere near the 10 000 ppm or more that now appears to be necessary.[7] Another difficult point in believing micellar solution to be important in primary migration is that the process of unloading the hydrocarbons carried by the fluid (water, micelles and hydrocarbons) at the final trapping position in the reservoir cannot be thoroughly explained.

The preceding discussion may lead us to conclude that the larger proportion of liquid hydrocarbons must migrate as a separate phase, although the rest can migrate in solution in water. In the case of gas migration, the proportion moving in solution in water can be relatively large, because of its greater solubility in water.

Although the form in which hydrocarbons moved at the time of primary migration, and the mechanisms of that migration, are not completely understood, the movement of water in fine-grained source rocks must be one of the most important factors. The amounts of organic matter and generated hydrocarbons in the source rocks are quite small in comparison with the amount of water. The movement of the large quantity of water must have influenced and may have controlled the direction and effectiveness of hydrocarbon migration.

Movement of Water in Sediments

There are essentially two different kinds of water moving in a sedimentary basin. Their respective characteristics are as follows.

Sediment Source Water

1. The movement of this type of water takes place in any part of a sedimentary basin (deep or shallow).
2. The principal direction of small-scale movement is from a shale or clay to a sandstone or other permeable bed.
3. The direction of large-scale movement is from the basin centre to its edges, or from the deeper parts to the shallower.
4. The amount of water is limited because the amount of sediment in a basin is limited.
5. Movement of this type of water is probably important in the primary migration of hydrocarbons.
6. Most movement of this type of water took place in the geological past.

Meteoric Water

1. The movement of this type of water is important in the relatively shallow intervals of a sedimentary basin.
2. The direction of small-scale movement can be either from sandstone to shale or from shale to sandstone. However, most movement of this type of water may take place in sandstones only.
3. The direction of large-scale water movement is from the basin edges to its centre, or from shallow to deep.

4.   There can be a very large amount of water.
5.   The movement of this type of water is probably unimportant in primary migration, but it may affect the trapping of hydrocarbons in a pool.
6.   Movement of this type of water is a present event, and may or may not have developed in the geological past.

Figure 2 shows a schematic diagram in which water is moving in an aquifer. The water pressure is measured at two points, A and B, in the

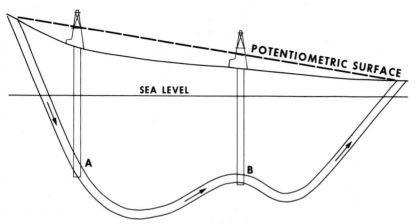

FIG. 2.   Schematic diagram showing water flow in aquifer due to hydrodynamic force.

aquifer. An imaginary vertical water column which corresponds to the measured pressure is shown, and the height of the column above the datum level (sea level in this case) is known. This is a measure of the potential level, and is called a potentiometric or piezometric surface. Water moves in the aquifer from a higher potential point (A) to a lower one (B) as shown in Fig. 2. In this case it is meteoric water.

If the datum level is taken at the surface, the potentiometric surface elevation can be shown as a function of excess pressure above hydrostatic pressure. Therefore, the excess pressure can also be used in determining the direction of fluid movement.

Figure 3 depicts a schematic example in which two aquifers, A and B, have different potential levels. If there is any fluid communication route between these aquifers, the fluid will move from the higher potential point to the lower (or from the higher excess-pressure point, A, to the lower, B).

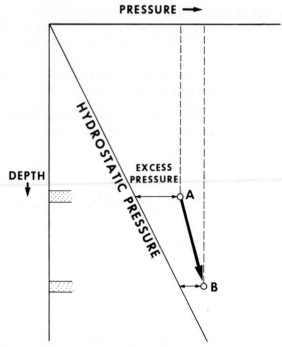

PRESSURE →

DEPTH

HYDROSTATIC PRESSURE

EXCESS PRESSURE

A

B

FIG. 3. Schematic pressure–depth plot for two aquifers, A and B. Arrow shows possible fluid-flow direction.

Note that the total fluid pressure at B is greater because of its greater depth, but its potential or excess pressure is lower, so that the fluid moves towards B from A.

The excess-pressure difference discussed above can be caused by the difference in elevations of the water-intake areas of the aquifers, if the water is meteoric water. If, however, the moving fluid originated in the sediments, loading of the sediment layers would be the principal cause of the excess fluid pressure.

Compaction Fluid Movement

Consider a clay or shale sequence in which clays or shales have reached compaction equilibrium and within which the fluid pressure is hydrostatic (stage A of Fig. 4). In other words, there is no excess fluid pressure at this stage. A thin sediment layer whose thickness is l_0 is deposited under water in a unit time interval. If the entire shale sequence reaches a new

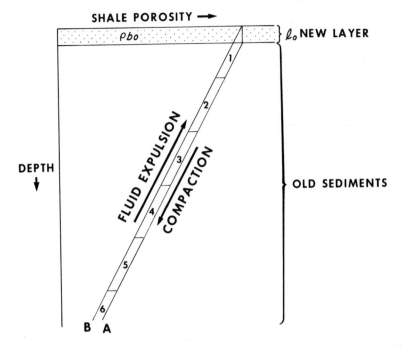

FIG. 4. Schematic diagram showing shale porosity–depth relation before and after deposition of uppermost sediments of thickness l_0, and the concepts of stepwise compaction and water expulsion.

equilibrium condition of compaction after deposition of this layer, a porosity–depth relationship such as is shown by stage B in Fig. 4 would be established. An exponential function between shale porosity and depth would be established at stage A or stage B (compaction equilibrium condition). Suppose there is an impermeable rock at the base of this sequence, so that the outlet for the expelled fluids exists only at the surface. The fluids move upwards. The compaction of the shales from stage A to stage B in this case would occur from the shallower part to the deeper part of the sequence. Figure 4 shows schematically that the porosity decrease occurs stepwise from the shallower part to the deeper (from 1 to 6)

To discuss fluid migration under these circumstances, it is necessary to determine the excess fluid pressure caused by the instantaneous deposition of the new layer l_0. The excess-pressure increase in this case is expressed as

$$E_l = (\rho_{b0} - \rho_w)gl_0 \qquad (1)$$

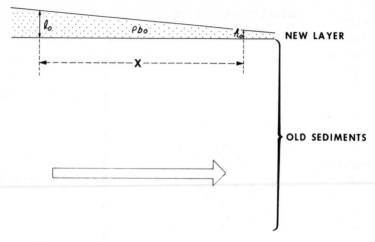

FIG. 5. Schematic diagram showing how thickness of uppermost sediments changes from l_0 to h_0 within distance X. Arrow shows compaction fluid-flow direction.

where E_l = excess fluid-pressure increase, ρ_{b0} = density of newly added sediment, ρ_w = density of water, l_0 = thickness of newly added sediment, and g = acceleration due to gravity.

When the thickness of the newly deposited sediment changes, the excess-pressure increase will change accordingly (Fig. 5). The excess-pressure increase E_h due to the sediment thickness h_0 is similarly shown as

$$E_h = (\rho_{b0} - \rho_w)gh_0 \tag{2}$$

If the distance between the points of different thicknesses, l_0 and h_0, is X, the horizontal excess-pressure gradient $(dp/dZ)_H$ due to the new loading of the wedge-shaped sediment is given as

$$\left(\frac{dp}{dZ}\right)_H = \frac{E_l - E_h}{X} = (\rho_{b0} - \rho_w)g\left(\frac{l_0 - h_0}{X}\right) \tag{3}$$

where $(l_0 - h_0)/X$ is considered to be the rate of thickness change of the new sediment with distance. The direction of the horizontal fluid movement in this case is from the thicker bed to the thinner (Fig. 5).

The vertical excess-pressure gradient $(dp/dZ)_V$ is expressed as follows:[26]

$$\left(\frac{dp}{dZ}\right)_V = \frac{dE_l}{dl_0} = \frac{dE_h}{dh_0} = (\rho_{b0} - \rho_w)g \tag{4}$$

If compaction and fluid expulsion are assumed to take place stepwise from surface to depth, the excess-pressure gradient given by eqn. (4) will be carried downward from the shallower depths to the deeper until a new compaction equilibrium condition B is completed for the entire sequence.

By comparing eqn. (3) with eqn. (4), we recognise that the horizontal excess-pressure gradient is much less than the vertical, because in most sedimentary basins the value $(l_0 - h_0)$ X tends to be quite small.

The range of the value $(l_0 - h_0)$ X in sedimentary basins may be guessed from regional geological cross-sections in the Gulf Coast and western Canada basins (Figs. 6 and 7). The values for basins experiencing rapid deposition generally would be greater than those for basins in which deposition was slower. Accordingly, in the Gulf Coast basin—a typical example of rapid deposition—the value for the Tertiary is about 1/40; in the western Canada basin it is about 1/250 for Cretaceous sediments and even less for older rocks ($\simeq 1/400$). Indeed, as these values refer to compacted sediments, the values for sediments that are being deposited may be assumed to be roughly twice as large if the effect of sediment compaction is allowed for. In other words, in most sedimentary basins the value $(l_0 - h_0)$ X for new sediments ranges from about 1/20 to 1/200. These values are relatively very small. Thus, the horizontal excess-pressure gradient in most sedimentary basins is 1/20 to 1/200 of the vertical excess-pressure gradient (see eqns. (3) and (4)).

The preceding paragraphs discussed the directions of horizontal and vertical fluid movements during sedimentation and the excess-pressure gradients caused by sediment loading. An important assumption in this migration model is that there is an outlet for water in the upward or horizontal direction. The next question is, what amounts of fluids have moved horizontally and vertically? Before approaching this problem, however, we must determine the total water loss from sediments due to compaction.

The relationship between the volumes of sediments before and after compaction is given as follows:[32]

$$V_0(1 - \phi_0) = V(1 - \phi) \qquad (5)$$

where V_0 = volume before compaction, V = volume after compaction, ϕ_0 = porosity before compaction, and ϕ = porosity after compaction. This equation assumes that the sediment matrix has suffered no mineralogical change during compaction, and that the porosity reduction is due entirely to fluid expulsion and compaction. The volume of the total

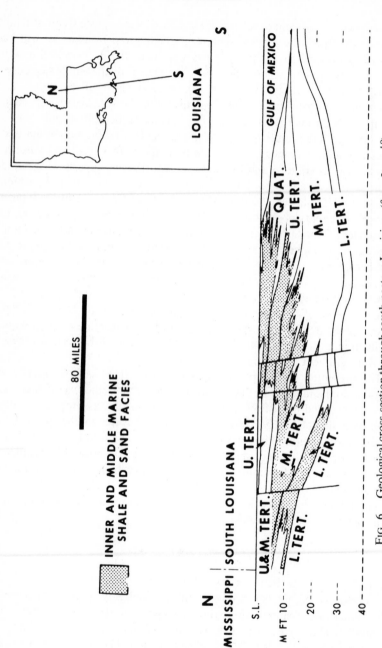

FIG. 6.   Geological cross-section through south-eastern Louisiana (from Jones[19]).

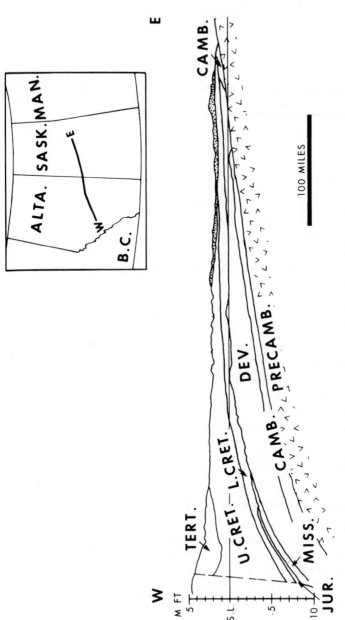

Fig. 7.   Geological cross-section in western Canada (from Gussow[14]).

fluid loss, W, due to compaction equals the difference between the volume of sediments before and that after compaction, or

$$W = V_0 - V = V\left(\frac{\phi_0 - \phi}{1 - \phi_0}\right) \tag{6}$$

To calculate the proportions of horizontal and vertical fluid movements, Darcy's equation may be used:

$$q_H = -\frac{k_H}{\mu}\left(\frac{dp}{dZ}\right)_H = -\frac{k_H}{\mu}(\rho_{b0} - \rho_w)g\frac{l_0 - h_0}{X} \tag{7}$$

$$q_V = -\frac{k_V}{\mu}\left(\frac{dp}{dZ}\right)_V = -\frac{k_V}{\mu}(\rho_{b0} - \rho_w)g \tag{8}$$

(see eqns. (3) and (4)) where q = volume of fluid moving through the sediments per unit area and unit time, k = permeability of the sediments, μ = viscosity of the fluid, and H and V as subscripts denote the horizontal and vertical directions of movement. Dividing eqn. (7) by eqn. (8), we obtain:

$$\frac{q_H}{q_V} = \frac{k_H}{k_V}\frac{l_0 - h_0}{X} \tag{9}$$

The validity of applying Darcy's equation to fluid movement in a shale sequence may be a matter for discussion. There is an opinion that Darcy's equation does not represent the fluid-flow situation in shales. However, extensive studies of under-compacted shales and abnormal pressures in the young sedimentary basins of the world suggest that the absence of permeable beds (e.g. sandstones) is probably the most important factor in causing these under-compacted shales, which have resulted from subnormal fluid expulsion.[12] If they are interbedded with many permeable sandstones of large areal extent, the shales will lose more fluids and compact to a near-normal value.

Although we do not know the exact mechanism of fluid migration in shales, such migration appears to be influenced or possibly even controlled by the mechanism of fluid movement in the interbedded permeable rocks. Darcy's equation is known to be applicable to such permeable rocks.

When applying Darcy's equation to a sandstone–shale sequence or possibly a shale sequence, however, we would have to vary the fluid viscosity, μ, with compaction. Figure 8 shows the result of estimates by Low[23] of water viscosity in montmorillonite. The viscosity (η) changes from about 1 to 8 cP. This finding suggests that when using eqn. (7) or eqn.

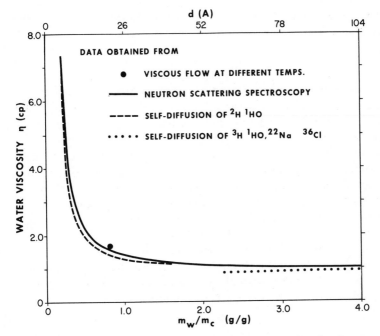

FIG. 8. Relationship between viscosity of water in clay and ratio of amount of water (m_w) over amount of clay (m_c) for distance (d) from clay surface (derived from Low[23]).

(8) to estimate the volume of fluid movement, one must increase the viscosity as compaction progresses. Permeability will, of course, decrease at the same time. In eqn. (9), however, which gives the ratio of the horizontal and vertical fluid volumes, the viscosity term is not included, so that the calculation is simpler.

The term $(l_0 - h_0)/X$ in eqn. (9) is the rate of thickness change of the new sediment layer, and k_H and k_V are the permeabilities of an older and deeper interval where fluid movements are being considered. The terms q_H and q_V are, of course, the volumes of fluids moving through the older and deeper interval during loading of the new sediment layer. Because the values k_H and k_V vary with burial and compaction, the ratio q_H/q_V changes with geological time even where $(l_0 - h_0)/X$ stays constant. If the ratio k_H/k_V stays constant throughout geological time, although k_H and k_V may vary, the ratio q_H/q_V is essentially controlled by $(l_0 - h_0)/X$.

The value $(l_0 - h_0)/X$ cannot be obtained directly from a geological section or an isopach map, because the sediments have already been compacted. Equation (9) can be converted to fit the rate of thickness change after compaction (or the rate of present thickness change) as follows (see Appendix):

$$\frac{q_H}{q_V} = \frac{k_H}{k_V}\left(\frac{l-h}{X}\right)\left(\frac{\rho_b - \rho_w}{\rho_{b0} - \rho_w}\right) \tag{10}$$

where $(l - h)/X$ is the rate of the sediments' thickness change with distance at present, or after compaction, and ρ_b is the bulk density of the sediments after compaction.

In the discussions above, we have noticed the importance of the two permeabilities—horizontal and vertical—in fluid migration associated with a wedge-shaped mass of sediments. Actually, in most sedimentary rocks, especially shales or siltstones, these permeabilities are seldom obtained. Therefore, it may be quite difficult to apply eqn. (10) to the real subsurface rocks. This problem may be simplified as follows.

Suppose there is an interbedded sandstone–shale sequence. If fluids were to move vertically upwards through this sequence, they would have to pass through both the sandstones and the shales, but the vertical flow rate would be largely controlled by the low-permeability shales, or

$$k_V \simeq k_{SH} \tag{11}$$

where k_{SH} is the permeability of the shales (in a strict sense, in the vertical direction), and as such could replace the term k_V in eqn. (10).

If the fluids were to move horizontally, they would move through the high-permeability sandstones because it was easier, and horizontal movement through the shales would thus be negligible. Therefore, the sandstone permeability becomes important. There must, of course, be a fluid outlet at the end of the sandstone. Fluid flow is also controlled by the sandstone thickness. If we consider a unit thickness of a sandstone–shale sequence, the sandstone thickness can be expressed in terms of the sandstone fraction, or percentage, in the unit sequence. As a result, k_H may be shown as

$$k_H \simeq \frac{S}{100}k_{SS} \tag{12}$$

where S is the percentage of sandstones in the unit sequence and k_{SS} is the sandstone permeability (in the horizontal direction).

By replacing k_H and k_V in eqn. (10) with k_{SS} and k_{SH}, using eqns. (11) and (12), we obtain

$$\frac{q_H}{q_V} = \frac{S}{100}\frac{k_{SS}}{k_{SH}}\left(\frac{l-h}{X}\right)\left(\frac{\rho_b - \rho_w}{\rho_{b0} - \rho_w}\right) \tag{13}$$

An interesting fact in eqn. (13) is that, if S or k_{SS} is very small, the ratio q_H/q_V becomes very small because the values $(l - h)/X$ in most sedimentary basins are relatively very small. In other words, the volume of the horizontal fluid movement relative to the vertical is quite small. This means that, if the shales are thick and homogeneous, most fluids will move vertically. The presence of some contiguous or lenticular sandstones in a thick shale sequence may not drastically change this basic direction of fluid flow.

Price[36] recently proposed the importance of growth faults through massive and under-compacted shales in the Gulf Coast as the main fluid-migration pathways. He suggested that hydrocarbons generated in the deep and hot under-compacted shale section have migrated upwards along these faults, in the form of a molecular solution in water. The current fluid-flow model, however, indicates that the principal direction of flow through massive shales is vertically upwards, whether the shales are faulted or not: in other words, such upward movement of generated hydrocarbons through these shales is always possible. Its importance in the total petroleum accumulations in this area, however, may not be so great because the total volume of vertical fluid flow through these under-compacted shales may not have been large.

If shales are interbedded with comparatively thick and permeable sandstones (S and k_{SS} in eqn. (13) are relatively very large), q_H/q_V could become relatively large, resulting in a predominantly horizontal fluid flow. The permeability of most reservoir sandstones probably exceeds 50 md, and that of shales is generally less than 0·05 md.[26] Therefore, the ratio k_{SS}/k_{SH} may be estimated to be greater than 1000 for most sandstone–shale sequences. Combining this figure, other assumptions (say, $S = 50\%$ and $[(\rho_b - \rho_w)/(\rho_{b0} - \rho_w)] = 2$), and the value $(l - h)/X$ in most sedimentary basins discussed previously, we conclude that in sandstone–shale interbeds q_H/q_V would be greater than 1; more fluids would move laterally than vertically. This type of fluid migration can be very important in many structures where the interval in the syncline that is stratigraphically equivalent to the main reservoir section at the crest has reached a temperature high enough to yield hydrocarbons. The hydrocarbons may easily move laterally along the bedding planes to the structure.

If the values of S and k_{SS}/k_{SH} are constant, on the other hand, q_H/q_V increases as $(l - h)/X$ increases. In other words, in a basin of rapid deposition such as the Gulf Coast, where $(l - h)/X$ is relatively large, there will be more horizontal fluid flow, other conditions being equal.

If the total fluid loss, W, from a given (present) volume of rock is calculated by eqn. (6), the horizontal, W_H, and the vertical, W_V, portions of the total fluid loss, W, can be estimated as follows:

$$W_H = \frac{q_H}{q_H + q_V} W = \left(\frac{q_H/q_V}{q_H/q_V + 1} \right) W \tag{14}$$

$$W_V = W - W_H = \left(\frac{1}{q_H/q_V + 1} \right) W \tag{15}$$

where q_H/q_V can be calculated from eqn. (13).

Now let us calculate the proportions of horizontal and vertical fluid volumes when a block of rock has lost 100 cubic feet of fluid during a given geological period; such an estimate can be made by using eqn. (6). Other factors assumed are as follows:

$$\frac{l - h}{X} = \frac{1}{100}$$

$$\frac{\rho_b - \rho_w}{\rho_{b0} - \rho_w} = 2 \cdot 0$$

$$\frac{k_{SS}}{k_{SH}} = 1000$$

$$\frac{S}{100} = 0 \cdot 3 \ (\text{or } S = 30\%)$$

The answer is as follows:

$$\frac{q_H}{q_V} = 0 \cdot 3 \times 1000 \times \frac{1}{100} \times 2 = 6$$

$$W_H = \left(\frac{6}{6 + 1} \right) \times 100 = 86 \text{ cubic feet}$$

$$W_V = 100 - 86 = 14 \text{ cubic feet}$$

In other words, in this case most of the 100 cubic feet of fluid moved horizontally.

The model discussed is applicable if the sediments reached the compaction equilibrium condition after each increment of instantaneous loading. If, however, some shales were to stay (slightly) under-compacted while other shales attained almost compaction equilibrium, significant pressure differences and barriers within the shale zones would be developed. This type of facies was named 'Mixed compaction facies'.[11] It occurs in an intermediate depth range below the normal compaction facies. An example of the calculated fluid-pressure profile of the mixed compaction facies in the Beaufort Basin, Canada, is shown in Fig. 9. Fluid moves from a higher excess-pressure point to one that is lower, and the inferred directions of fluid flow are shown by arrows. Similar pressure or compaction patterns were reported in other sedimentary basins.[24,25,27]

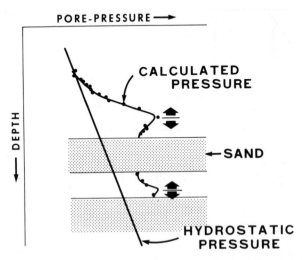

FIG. 9.   Example of calculated fluid pressure–depth plot of a Beaufort well, Canada.

Once an interbedded sequence has reached such an intermediate compaction stage, essentially all the compaction fluids may have to move laterally through the interbedded sandstones. There is some vertical fluid flow in the shales, too, but the flow is only local. In summary, the development of the mixed compaction facies could facilitate lateral fluid flow from synclinal areas, and this flow would take place after the sandstone–shale sequence has reached an intermediate depth range where petroleum may have been generated by the thermal process.

The preceding discussion introduced the method of calculating the horizontal and vertical fluid volumes that have moved from a given block of rock. In the subsurface, however, fluids expelled from the other blocks below and beside a particular block will also influence the fluid-flow condition within that block. In other words, the cumulative effect of fluid migration will be three-dimensional. Although estimating such a fluid-migration condition is extremely complicated, it may be worthwhile in that

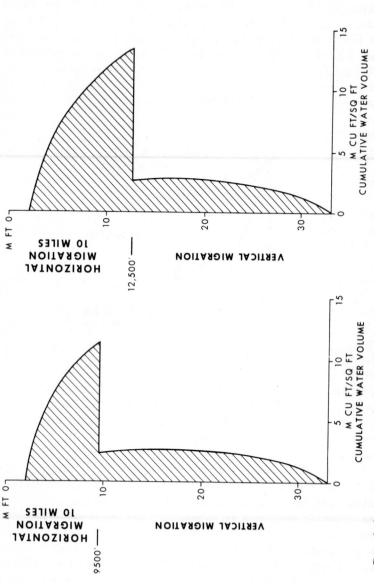

FIG. 10.   Cumulative water-loss volumes from shales in Gulf Coast (combined vertical and horizontal migration model). Reproduced from Ref. 32 by permission of the American Association of Petroleum Geologists.

the migration of hydrocarbons may be affected by the cumulative fluid migration after the hydrocarbons have been generated.

Magara[32] recently estimated cumulative compaction-fluid volumes, using Dickinson's[8] porosity–depth curve and a simplified Gulf Coast model. In this model, the upper sequence is composed of sandstone–shale interbeds in which fluids have moved horizontally, and the lower sequence consists of massive and homogeneous shales where the compaction fluid has moved vertically upwards. The horizontal migration distance in the upper sequence is assumed to be 10 miles, and the total thickness of the sedimentary column 33 000 ft (10 km). Figure 10 shows the cumulative volumes of fluid loss since burial to 2000 ft from a shale column whose base area is 1 sq. ft (the respective depths to the boundaries of the upper and lower sequences are assumed to be 9500 ft and 12 500 ft).

It is interesting to note that the cumulative fluid–volume plot based on the model that simulates the Gulf Coast sedimentary basin resembles the oil-production frequency plot for the same area.[5] This similarity suggests that fluid movement due to sediment compaction is one of the controlling factors in hydrocarbon occurrence in that area.

A fact that could affect the importance of mechanical shale compaction and fluid expulsion in petroleum migration is that the rate of compaction decreases continuously as the shales become more deeply buried. In other words, by the time the source rocks had reached deep burial where the temperature was high enough to generate hydrocarbons, the movement of compaction fluids might have become too slow and insignificant.

If fluids expand at such depths, the expansion might facilitate late-stage fluid movement. Subsurface temperature increase with burial depth might cause such fluid expansion in most sedimentary basins.

Aqua-thermal Fluid Movement

Figure 11 is a pressure–temperature diagram for water with selected iso-density lines, adapted from Barker.[3] The vertical scale is pressure in psi, and the horizontal scales are temperature in both centigrade and Fahrenheit. Density values in g/cm^3 (and specific volume values in cm^3/g) of water are shown along the iso-density lines. The original data for constructing this diagram were obtained by Kennedy and Holser.[20] The three geothermal-gradient lines of 25 °C/km (1·37 °F/100 ft), 18 °C/km (1 °F/100 ft) and 36 °C/km (2 °F/100 ft) for hydrostatically pressured water (the system is not closed) are superimposed; the lines intercept water iso-density lines whose values decrease as the pressure (or burial depth) increases. A hydrostatic pressure gradient of 0·47 psi/ft was used. This

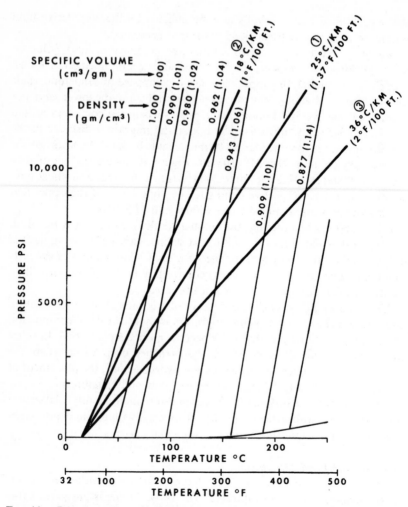

FIG. 11. Pressure–temperature–density (or specific volume) diagram for water. Three geothermal gradient lines of 25, 18 and 36°C/km for hydrostatically pressured fluid are superimposed on a basic diagram derived from Barker[3].

progression to lower densities and higher specific volumes means that a given weight of water expands with burial: the reason is that the increase of pressure associated with the 0·47 psi/ft hydrostatic gradient is inadequate to hold the water volume constant. The amount of expansion can be derived easily from the specific volume values (cm³/g), shown in brackets.

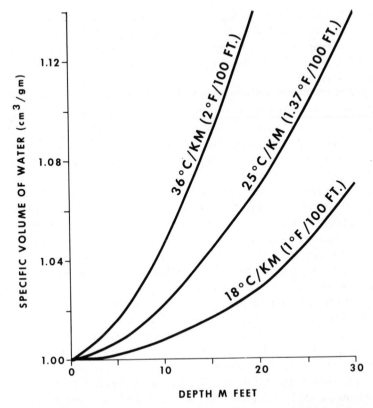

FIG. 12.   Specific volume (of water)–depth relations in normally pressured zones for three geothermal gradients of 25, 18 and 36 °C/km. Reproduced from Ref. 29 by permission of the American Association of Petroleum Geologists.

When the geothermal gradient is 25 °C/km (1·37 °F/100 ft), for example, the specific volume increases from 1 cm³/g at 0 psi pressure to 1·10 cm³/g at 11 600 psi, which corresponds to a burial depth of about 25 000 ft. Thus, a 10 % water expansion results from about 25 000 ft of burial; this is a significant amount.

Continuous expansion of water for the three geothermal gradients is depicted in Fig. 12, where the specific volume of water (cm³/g) is shown on the vertical scale and depth (ft) on the horizontal scale. At 20 000 ft, for example, about 3 % expansion has occurred for the geothermal gradient of 1 °F/100 ft, about 7 % expansion for 1·37 °F/100 ft, and 15 % for 2 °F/100 ft.

Figure 12 shows that rates of increase in specific volume, or rates of expansion, increase with burial depth. This fact is interesting because the amount of water expelled by compaction decreases with burial depth, but the subsurface temperature tends to expand the water volume. This expansion could facilitate fluid migration at depth and hence could favour hydrocarbon migration.

Expansion of rock grains also may be considered in the discussion of fluid migration. The grain expansion would create more inter-grain space; thus more space for water. Its effect, however, is much less significant; the thermal expansion of quartz, for example, is only about 1/15 that of water.[39] Thermal expansion data for a dry clay matrix are not readily available; the value for quartz may be the closest approximation. In other words, if the ratio of volume of water to that of rock grains is more than about 1:15 (porosity is more than about 6%), the effect of water expansion overrides that of grain expansion, resulting in water movement. In the Gulf Coast, a shale porosity of 6% would not be attained above 24 000 ft.[8]

Note that the above-mentioned aqua-thermal model is valid when pore water is not completely isolated. Such a relatively open system is developed in the normal and mixed compaction facies, which usually occur in the shallow to intermediate depth range in many sedimentary basins. If the pore fluids are more isolated, as in the case of under-compacted facies, the fluid cannot expand freely and the fluid pressure will increase.[3,29,31]

The directions of fluid migration due to the aqua-thermal effect are from a hot place to a cold, from a deep section to a shallow, and from a basin centre to its edges. These directions are essentially the same as those of fluid movement caused by sediment compaction. Therefore, the significance of the aqua-thermal effect in the subsurface may simply be to increase the effectiveness of compaction fluid flow at deep burial.

Now let us assume a geological model at intermediate depths in which sandstones are interbedded with shales. A shale–porosity profile such as is shown in Fig. 13A may be developed. If the interbedded sandstones are permeable, the maximum fluid expulsion or the maximum shale-porosity reduction will occur in the shales directly above and below the sandstones. The porosity in the middle of a shale bed may remain relatively high. The corresponding fluid-pressure plot is shown in Fig. 13B, in which arrows depict the inferred directions of compaction fluid flow.

If water expands from the thermal effect, water will move within the shale bed from the centre to the upper and lower edges, because more expansion can be expected at the point of higher porosity (more water). The directions of the small-scale fluid migration due to the aqua-thermal effect are essentially the same as those of compaction fluid migration.

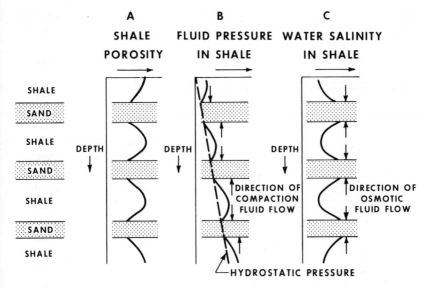

FIG. 13. Schematic diagram showing shale porosity, fluid pressure and pore-water salinity distributions in an interbedded sand–shale sequence. Reproduced from Ref. 28 by permission of the American Association of Petroleum Geologists.

Osmotic Fluid Movement

In many sedimentary basins the salinity of the formation water increases with depth or compaction. These salinity values are usually higher than that of sea-water (about 35 000 ppm). In the under-compacted zones, the salinity is lower than those of normal and mixed compaction zones. The main cause of these salinity variations in sedimentary rocks may be ion-filtration by shales.[28]

Ion-filtration by clays or shales has also been demonstrated by laboratory methods,[10,34] which showed that the clays and shales filter salt from a solution. Therefore, the fluids moving through the shales must be fresher than the original solution that saturated the shales.

Hedberg[15] studied pore-water chlorinities and porosities in shales, using cores from several areas in the world. Figure 14 shows the chloride content (ppm) versus porosity plots for the Burgan field in Kuwait and three oil fields in Texas. The relation between the chlorinity† and porosity in the Burgan data may be approximated by a hyperbola; the chlorinity increases as the porosity decreases. The data from the three Texas fields are too

† Salinity (NaCl) may be calculated by multiplying the chlorinity by 1·65.

scattered and insufficient to prove or disprove the hyperbola relation. It is, however, interesting that most of the plotted data from Texas fall within the extension of the general Burgan trend.

Combining the concept of ion-filtration and the shale-porosity profile as shown in Fig. 13A enables a possible water-salinity profile for the shales to

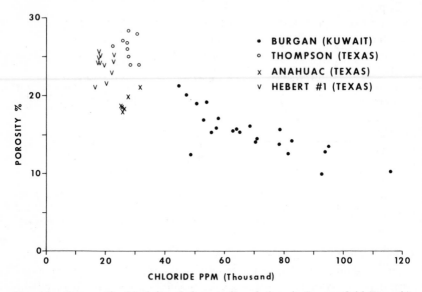

FIG. 14. Pore-water chlorinity–shale porosity relations in Burgan field (Kuwait) and three fields in Texas. Reproduced from Ref. 28 by permission of the American Association of Petroleum Geologists. Original data derived from Hedberg.[15]

be drawn (Fig. 13C). The salinity varies inversely with shale porosity, i.e. it increases as the porosity decreases. Salinity, therefore, would increase from the centre to the edges of each shale bed. Because osmosis tends to move water from a fresher to a more concentrated site, the fluid-flow direction due to osmosis can be inferred to be as shown by the arrows in Fig. 13C.

The osmotic-pressure difference due to salinity change is probably not very large as compared with that due to compaction. According to the chart shown by Jones,[19] the osmotic-pressure difference caused by a salinity difference of 50 000 mg/l is only about 600 psi. Because osmotic fluid flow is in the same direction as compaction fluid flow, however, osmotic flow could facilitate the primary hydrocarbon migration from shales to permeable sandstones.

Fluid Movement due to Clay Dehydration

Powers[35] showed that alteration of montmorillonite to illite in the Gulf Coast area begins at a depth of about 6000 ft and continues at an increasing rate to a depth, usually about 9000–12 000 ft, where there is no montmorillonite left. This alteration offers a mechanism for desorbing the last few layers of adsorbed or bound water in clay and transferring it into inter-particle locations as free water. If the last few layers of bound water have a greater density than free water, this released water tends to increase its volume as it is desorbed from between unit layers. If water expansion is restricted, the pore-water pressure will increase to abnormally high levels.

According to Burst,[5] clay dehydration depends mainly on subsurface temperature, the average dehydration temperature being 221 °F in the Gulf Coast. Certain chemical conditions for potassium fixation also are required for this conversion. Phase change and possible expansion of bound water at the time of dehydration may, as proposed by Burst,[5] be important agents for flushing hydrocarbons, at least from clay-interlayer locations to inter-particle locations (shale pore space).

Martin[33] summarised data on adsorbed water density in montmorillonite analysed by several different investigators. This summary is shown in Fig. 15, which plots the calculated and measured water density versus amount of water in the clay ($g\,H_2O/g\,clay$). Figure 15 appears to support Powers'[35] and Burst's[5] proposals of the higher-than-normal (greater than $1\,g/cm^3$) density of the adsorbed water. However, the validity of the higher-than-normal density shown in Fig. 15 is not very certain, because most of these higher values were derived from calculations rather than direct measurements. Martin[33] stated that the only unambiguous adsorbed water-density data are [those] of Anderson and Low,[1] which show values less than $1\,g/cm^3$.

Therefore, from the data existing at present, it is difficult to prove or disprove the water expansion and flushing effect associated with clay dehydration. However, we may be able to say that clay dehydration could be an additional source of liquid water at relatively deep burial where hydrocarbons may have been generated.

Van Olphen[41] demonstrated that at 25 °C the pressure needed to remove the last interlayer of water is 65 000–70 000 psi, and that needed for the second-to-last water interlayer is 30 000 psi. These values are considerably higher than the pressure at depths less than 20 000 ft. Overburden pressure alone, then, may not suffice to release at least the last two layers of bound water. This is the main reason why Burst[5] and Powers[35] developed their

KINJI MAGARA

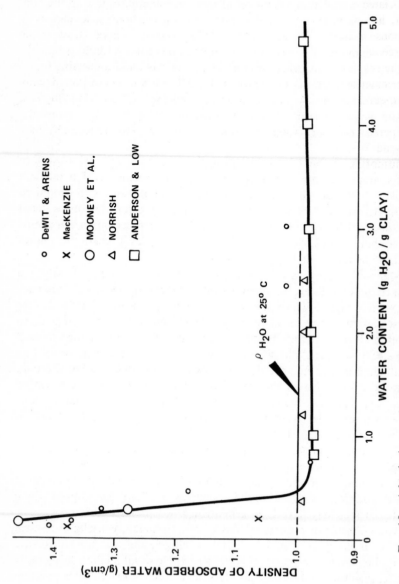

FIG. 15. Adsorbed water density on Na-montmorillonite. Reproduced from Ref. 33 with permission.

concepts of the temperature-dependent water-release mechanism associated with clay mineral conversion.

If, however, the interlayer water is released by clay dehydration in response to temperature rise, and subsequently remains in the pore spaces as free water, the same overburden pressure could be enough to push it out of the shales, provided drainage is available. This type of possible water movement is essentially the same as that caused by sediment compaction discussed above.[30]

The validity of the average dehydration and mineral-conversion temperature of 221°F proposed by Burst[5] must also be examined. Schmidt[38] studied the proportions of expandable clay (mostly montmorillonite) and non-expandable clay in a well drilled in the Gulf

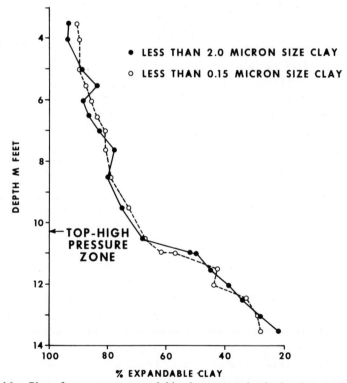

FIG. 16. Plot of percentage expandable clay versus depth showing accelerated increase in diagenesis of montmorillonite to mixed-layer montmorillonite–illite. Reproduced from Ref. 38 by permission of the American Association of Petroleum Geologists.

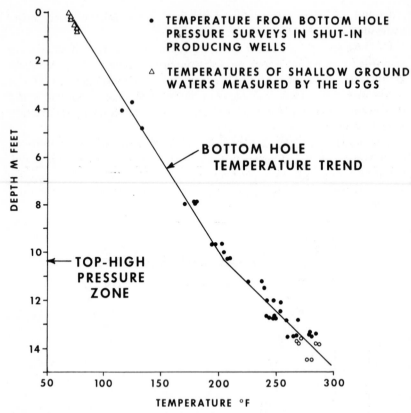

FIG. 17. Plot of measured temperatures versus depth, which shows increased temperature gradient at top of high-pressure zone. Reproduced from Ref. 38 by permission of the American Association of Petroleum Geologists.

Coast (Fig. 16). This figure shows that the rate of mineral conversion increases at about 10 500 ft, which corresponds to a subsurface temperature of about 200 °F (Fig. 17). However, the geothermal gradient in this well also increases at that depth (10 500 ft), which is the top of the under-compacted (abnormally-pressured) section. Because water has a thermal conductivity significantly lower than that of most rock matrix, the under-compacted section, which contains an excess amount of water, tends to have a thermal conductivity lower than that in the normally-compacted section.[22] If heat flows upwards at a given rate, the geothermal gradient in the under-compacted section would become greater than that in the normally-compacted section.

An important point shown in Fig. 16 is that clay mineral conversion is not a drastic event. The conversion temperature of 221 °F, as suggested by Burst,[5] may not be required. Rather, conversion seems to begin almost immediately after deposition, and continues to depth. The higher the geothermal gradient, the faster the rate of conversion.

Because in essentially all the world's sedimentary basins temperature tends to increase with depth, the bound water will be released in any case. Clay mineral conversion could create an additional source of liquid water at depth. Its significance in primary migration, however, cannot be understood clearly, because whether such conversion causes fluid expansion and migration is not known.

Other Possible Causes of Primary Migration
There are several other possible causes of primary migration, such as capillary pressure, buoyancy, diffusion and generation of hydrocarbons—especially gas, etc. These causes are mostly unassociated with the movement of water. Although there is no solid reason to deny their importance, I personally feel that water movement of some kind must be important at the primary-migration stage, and that therefore a mechanism unrelated to water movement may be of secondary importance. As pointed out previously, we are dealing with a large amount of water and a relatively small amount of hydrocarbons in the sediments, which have a very fine network at the time of primary migration.

Form of Hydrocarbons in Primary Migration
If all the hydrocarbons are in molecular solution in water at the primary-migration stage, estimating the volume and direction of flow of sediment source water as discussed is of prime importance in understanding hydrocarbon migration. The water volume may be tied directly to the amount of hydrocarbons. However, the solubility of liquid hydrocarbons in water is relatively low, even at elevated temperatures (Fig. 1). Micellar solution as proposed by Baker[2] cannot be very important in the subsurface for the several reasons mentioned above. According to Dickey's[7] estimate, there must be at least 10 000 ppm hydrocarbons in the flowing stream at the time of primary migration.

Another approach to estimating the required concentration of oil in the flowing stream is as follows. Tissot and Pelet[40] analysed the amounts of hydrocarbons, resins and asphaltenes in shales underlying a reservoir in Algeria. Figure 18 shows the results of their analyses in mg/g organic carbon. Although the amounts of resins and asphaltenes in the shale

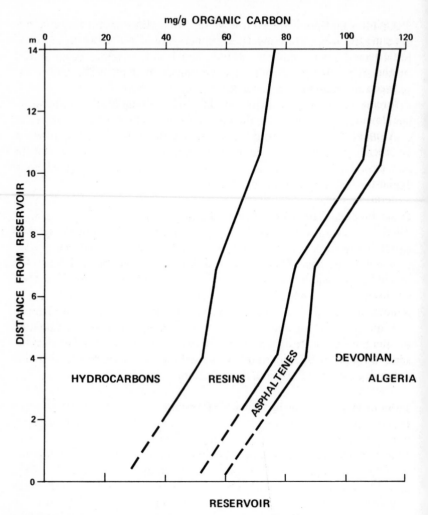

FIG. 18.   Plot of amounts of hydrocarbons, resins and asphaltenes versus organic carbon (mg/g) of Devonian shales underlying a reservoir in Algeria. Original data derived from Tissot and Pelet.[40]

remain relatively constant, the amount of hydrocarbons decreases toward the reservoir, suggesting primary hydrocarbon migration. The difference in hydrocarbon contents at the 14-m point and at the near-reservoir point is about 40 mg/g organic carbon. If the level of total hydrocarbon generation per gramme of organic carbon is constant throughout the shale section, this

40 mg represents the lowest possible amount of hydrocarbons expelled per gramme of organic carbon from the shale closest to the reservoir. If the shale has a density of 2 g/cm^3 and 1 weight percent of organic carbon, 1 cm^3 of this shale lost 0·8 mg of hydrocarbons.† If the porosity difference between these two points is 10%, which seems to be the largest porosity difference possible under these conditions, the amount of hydrocarbons in the flowing stream can be estimated to be about 8000 ppm.

As mentioned, this estimate is based on the lowest possible estimate of hydrocarbons in the compaction fluid; the true value could be higher. In any case, this figure is at least one order of magnitude higher than the highest molecular-solubility figure in the temperature range for oil generation, and is surprisingly close to the > 10 000 ppm given by Dickey.[7] Note that the density and porosity data for the shales studied by Tissot and Pelet[40] are not available, so that they have had to be assumed for this estimate.

Vyshemirsky et al.[42] experimented with squeezing a mixture of clay, liquid hydrocarbons and water up to 300 atm. They found that the amount of hydrocarbons squeezed with the water was more than could be accounted for by the solution mechanism alone.

From the above estimate and other observations, it is clear that the greater proportion of liquid hydrocarbons must move as a separate phase. Gas, however, can migrate in aqueous solution because of its higher solubility.

The question then arises: Why is the movement of water important if most of the liquid hydrocarbons move as a separate phase? The next section will suggest an answer.

Migration of Oil as an Oil Phase

A comprehensive discussion of the mechanisms associated with oil-phase migration was published in 1954 by Hobson.[16] Recently Dickey[7] rediscussed this possibility from a slightly different angle.

In compacted shales, the larger proportion of water is electrically charged at the clay surfaces, and has a relatively high viscosity (Fig. 8) which means that some water is semi-solid. The amount of liquid (or free) water in the compacted shales is probably not great. In these circumstances, if the shales compact further, the oil as well as the liquid water will migrate provided that the oil saturation in the liquid phase is higher than the critical value for oil migration.

† The figures used for this estimate would be the lowest possible values to produce the lowest possible hydrocarbon yield.

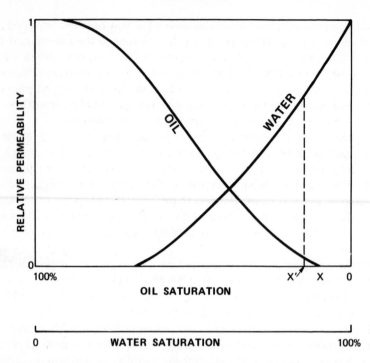

FIG. 19. Schematic diagram showing relative permeability–water (or oil) saturation relation for sandstone.

Basing his argument on the concept of relative permeability in sandstone, Dickey[7] stated that, 'oil will move along with water only if it occupies about 20 % or more of the pore volume.' If the sandstone is partly water-wet and partly oil-wet, the critical residual oil saturation can be as low as 10 %.[37] Dickey[7] also suggested that the residual oil saturation in shales may be less than 10 % and possibly as low as 1 %, because a considerable fraction of the internal surfaces of shales can be oil-wet.

A schematic diagram of the relative permeability and oil (or water) saturation relationship is shown in Fig. 19. The critical residual oil saturation is marked by an X, and for shales may be assumed to be a value between 10 and 1 %. For oil to migrate along with water, this critical oil saturation must be exceeded, that is if the oil saturation is at X' in Fig. 19, there will be some oil migration.

The amount of oil in a source rock is probably quite small. For the relative proportion of oil in the liquid phase to reach a point where it can

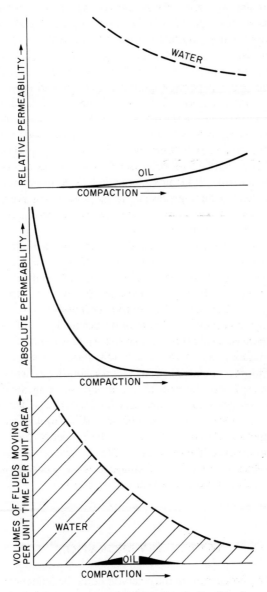

Fig. 20.   Schematic diagram showing a model of oil migration as an oil phase.

facilitate oil-phase migration, the amount of liquid water in the shale must always be relatively small. That is, both the original liquid water and the secondary liquid water generated from clay mineral conversion must be effectively drained from the shale. Such effective drainage is generally found in interbedded sandstone–shale intervals of shallow to intermediate depths.

If the liquid water generated cannot move out and stays in the shale pores, the oil concentration in the liquid phase will become less. This situation may be observed in the deep and under-compacted shales in many young sedimentary basins. Most of the semi-solid water in the montmorillonite has already been released to become liquid water, but the liquid water generated has not been expelled for lack of a good permeable zone. Some water will still be moving through these shales at an extremely slow rate, but effective oil migration in the oil phase is not likely because the oil saturation is so low. Even in this case, however, most gas and some oil may move in solution in water.

On the basis of the above discussion, I propose a model for primary migration in the oil phase, as shown in Fig. 20. The top diagram shows a schematic of relative permeability versus degree of compaction in a shale. As the shale compacts, the relative permeability to water decreases and that to oil increases. Although the relative permeability to oil increases drastically with compaction, the absolute permeability of the shale will continually decrease as the shale loses more liquid water and becomes more compacted (middle diagram of Fig. 20). Oil migration in the oil phase will reach its maximum at an intermediate compaction stage, then decline as the absolute permeability of the shale decreases (bottom diagram of Fig. 20). If this peak oil-migration stage is not very far from the peak oil-generation stage, we may be able to expect significant oil accumulation.

An important conclusion derived from the concepts discussed above is that effective drainage of fluids is essential to effective oil migration as an oil phase. The effectiveness of the drainage can be worked out from the calculated cumulative fluid-loss volumes or calculated pressure plots, as discussed.

SECONDARY MIGRATION

Movement of hydrocarbons in the reservoir could be influenced by some of the agents already discussed, especially compaction fluid and the aquathermal effect. Certain other factors, however, are also involved at the secondary-migration stage. Once hydrocarbons from source rocks enter reservoir rocks, they will encounter several different physical conditions which did not exist in the source rocks;

(1) larger pore spaces,
(2) fewer capillary restrictions,
(3) less semi-solid water, and
(4) less fluid pressure.

These new conditions help to enlarge and connect the hydrocarbon globules. The vertical connection of the globules will produce a significant buoyant force. The reduced capillary restrictions will allow this buoyancy to move the inter-connected globules to a higher structural position in the reservoir.

The critical height of oil column, Z_o, for oil migration in such a case is given as follows by Berg:[4]

$$Z_o = 2\gamma \left(\frac{1}{r_t} - \frac{1}{r_p} \right) \bigg/ g(\rho_w - \rho_o) \qquad (16)$$

where γ = interfacial tension between oil and water; r_t = pore-throat radius; r_p = pore radius; g = gravity acceleration; ρ_w = density of water; and ρ_o = density of oil. Berg considered a well-sorted, fine-grained sandstone with a porosity of 26%. Such a natural aggregate may approximate a rhombohedral packing of uniform spheres in which pore sizes are $0.154D$, $0.225D$ and $0.414D$, D being the sphere diameter.[13] Figure 21 shows the critical height of oil column as estimated by these assumptions: if the oil column exceeds this critical height, the oil will move; otherwise, it will stay. In this model, the reservoir rock, whose grain size, D, is 0.2 mm, is overlain by the same or a finer rock. The critical height of the oil column (vertical scale) is shown for a given grain size (horizontal scale) and given density difference ($\Delta\rho$) between water and oil. If, for example, the height of oil column exceeds 5 ft when the fluid-density difference is 0.2 and the reservoir grain size is 0.2 mm, the oil will move upwards within the reservoir. If the reservoir rock becomes finer upward, or is overlain by other finer rock, a taller oil column is required for upward migration.

Although buoyancy itself may suffice to move hydrocarbons in many reservoir rocks, other forces such as sediment compaction, the aqua-thermal effect and possibly clay dehydration may also influence secondary migration. In many areas, however, the directions of migration due to these different causes are the same as for that due to buoyancy—from a structurally deeper place to one that is shallower. If the directions of fluid flow resulting from these other causes are different from that of the flow

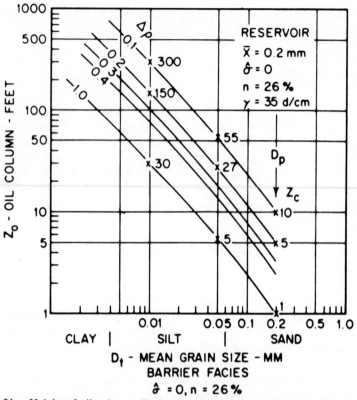

FIG. 21. Height of oil column, Z_o that can be trapped by barrier rock of mean grain size D_t in a reservoir rock of grain size $D_p = 0.2$ mm, where both rocks are composed of uniform spherical grains in rhombohedral packing and porosity, n, is 26%. Interfacial tension, γ, is assumed to be 35 dynes/cm and $\hat{\sigma}$ is the contact angle. Reproduced from Ref. 4 by permission of the American Association of Petroleum Geologists.

caused by buoyancy, hydrocarbon accumulations may be forced to move to locations other than the structural tops, or may be completely lost from the sedimentary basin. A good example of this type of accumulation is that resulting from the combination of structural and hydrodynamic conditions depicted by Fig. 22;[17] the accumulation is moved down-dip by the hydrodynamic force.

Whether the hydrocarbons that have moved to the structural top or to any other hydrodynamically-controlled position can be kept in the reservoir depends on the effectiveness of the seal. In the case of anticlinal or domal traps, the sealing capacity of the rock overlying the reservoir will

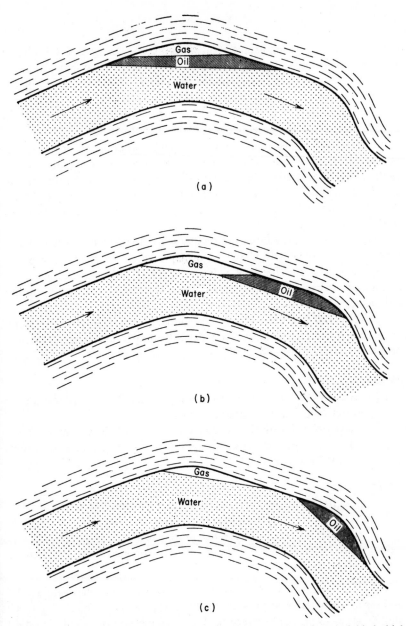

FIG. 22.   Types of hydrodynamic oil and gas accumulations in gently folded, thick sand. (a) Gas entirely underlain by oil; (b) gas partly underlain by oil; (c) gas and oil accumulations separated. Reproduced from Ref. 17 by permission of the American Association of Petroleum Geologists.

determine the maximum height of oil accumulation. In the case of fault and stratigraphic (pinch-out) traps, the sealing capacity of the material along the fault planes and the rocks beside the faults or pinch-outs will be the determining factor.

As mentioned earlier, Evans *et al.*[11] proposed the 'mixed compaction facies' in the Tertiary sequence of the Beaufort Basin. The mixed compaction facies is composed of slightly under-compacted (over-pressured) shales and normally-compacted shales and sandstones. These slightly over-pressured shales are considered to have restricted the vertical escape of the fluids in the sandstones and are called 'pressure seals'. About 90 % of the accumulated hydrocarbons (mostly gas) in this area are found in the mixed compaction facies.

Figure 23 shows an example of pressure seal in the Gulf Coast. The shales between 7000 and 9000 ft, which are over-pressured, overlie the normally-pressured sandstones between 9000 and 10 000 ft. These pressure seals commonly occur in an intermediate depth range in many young sedimentary basins.

Capillary seal has been known to engineers and geologists for many years, but pressure seal is relatively new. A comparison of the respective properties of pressure and capillary seals is of interest.

Pressure Seal
1.   Seal for any form of hydrocarbons whether they are in solution in water or occur as a separate phase.
2.   Developed during the intermediate stages of shale compaction.
3.   May be more important for gas than for oil, since gas is more soluble in water.

Capillary Seal
1.   Seal for hydrocarbons only as a hydrocarbon phase.
2.   May become more important during the later stages of shale compaction, when it becomes more effective.
3.   May be more important for oil than for gas, since gas is more soluble in water.

The effectiveness of pressure seals in the subsurface may be evaluated by making a calculated fluid-pressure profile of the overlying shales. Figures 9 and 23 are examples of pressure profiles. The method of calculating fluid pressures from well log data was discussed by Magara.[25,27] For a pressure seal to be effective, the excess pressure due to the buoyancy of the hydrocarbon column must be less than the excess pressure of the overlying

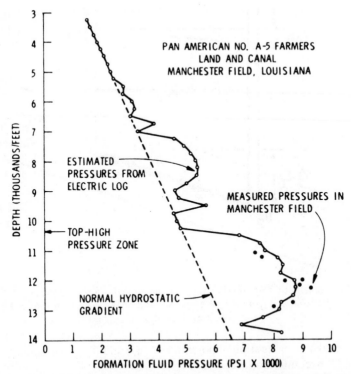

FIG. 23. Estimated formation fluid pressures from electric log derived shale properties compared with measured pressures from Manchester field, Louisiana. Reproduced from Ref. 38 by permission of the American Association of Petroleum Geologists.

shale above hydrostatic pressure. As pressure seals are usually associated with capillary seals, the combined effect may control the trapping condition.

Capillary sealing capacity may be evaluated by using Fig. 21 as by Berg.[4] For example, 100 ft of gas column (the density difference is assumed to be 1·0) may be retained by a capillary seal composed of clays of 0·003 mm grain size.

If water is moving as a result of hydrodynamic force in the reservoir, the critical height of oil column, Z_{ot}, can be expressed differently:[4]

$$Z_{ot} = \frac{2\gamma\left(\dfrac{1}{r_t} - \dfrac{1}{r_p}\right)}{g(\rho_w - \rho_o)} \pm \left(\frac{\rho_w}{\rho_w - \rho_o}\right)\frac{\mathrm{d}h}{\mathrm{d}X}X_o \qquad (17)$$

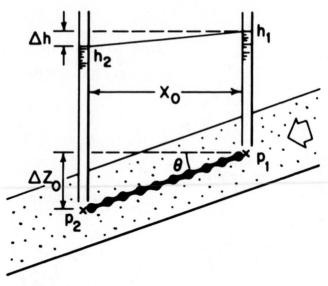

FIG. 24.   Schematic diagram showing oil stringer held in aquifer by down-dip flow of water. Reproduced from Ref. 4 by permission of the American Association of Petroleum Geologists.

where dh/dX = inclination of potential surface and X_o = horizontal width of the oil accumulation (see Fig. 24).

The optional sign in eqn. (17) refers to flow directions: the positive sign corresponds to down-dip flow and the negative to up-dip flow. In other words, more of a hydrocarbon column can be retained if there is a down-dip fluid flow or down-dip potential gradient, other conditions being equal.

The importance of structural timing, whether relatively early or late, in major hydrocarbon accumulation is a matter for controversy. Many geochemists think that late structure should not preclude the formation of a good accumulation, because petroleum maturation is a relatively late event, which requires geological time and temperature. On the other hand, however, many experienced geologists believe that relatively early structures offer better prospects for finding a major petroleum accumulation.[43] Although the true answer is not yet known, the fact that geochemical source-rock analysis is based on the hydrocarbons still remaining in the source rocks, rather than those that have already migrated, should not be overlooked. I myself feel that hydrocarbons that have migrated to reservoirs may have been matured sooner than most geochemical analyses suggest.

ACKNOWLEDGEMENTS

I am indebted to Imperial Oil Ltd, Canada, for permission to publish this paper, and to Dr G. D. Hobson of V. C. Illing and Partners and Dr R. P. Glaister of Imperial Oil Ltd for important comments.

APPENDIX

Applying the concept shown by eqn. (5), we obtain

$$l_0(1 - \phi_0) = l(1 - \phi) \tag{a}$$

$$h_0(1 - \phi_0) = h(1 - \phi) \tag{b}$$

The porosity can be expressed in terms of density as follows:

$$\phi_0 = \frac{\rho_m - \rho_{b0}}{\rho_m - \rho_w} \tag{c}$$

$$\phi = \frac{\rho_m - \rho_b}{\rho_m - \rho_w} \tag{d}$$

where ρ_b = bulk density of sediments after compaction, and ρ_m = matrix (or grain) density of sediments. Introducing eqns. (c) and (d) into (a) and (b), we obtain

$$l_0 = l\left(\frac{\rho_b - \rho_w}{\rho_{b0} - \rho_w}\right) \tag{e}$$

$$h_0 = h\left(\frac{\rho_b - \rho_w}{\rho_{b0} - \rho_w}\right) \tag{f}$$

Introducing eqns. (e) and (f) into eqn. (9) results in

$$\frac{q_H}{q_V} = \frac{k_H}{k_V}\left(\frac{l - h}{X}\right)\left(\frac{\rho_b - \rho_w}{\rho_{b0} - \rho_w}\right) \tag{10}$$

REFERENCES

1. ANDERSON, D. M. and LOW, P. F. (1958). The density of water adsorbed by lithium-, sodium-, and potassium-bentonite, *Soil Sci. Soc. Am. Proc.*, **22**, 99–103.
2. BAKER, E. G. (1962). Distribution of hydrocarbons in petroleum, *Am. Assoc. Petrol. Geologists Bull.*, **46**, 76–84.

3. BARKER, C. (1972). Aquathermal pressuring: role of temperature in development of abnormal-pressure zones, *Am. Assoc. Petrol. Geologists Bull.*, **56**, 2068–71.

4. BERG, R. R. (1975). Capillary pressures in stratigraphic traps, *Am. Assoc. Petrol. Geologists Bull.*, **59**, 939–56.

5. BURST, J. F. (1969). Diagenesis of Gulf Coast clayey sediments and its possible relation to petroleum migration, *Am. Assoc. Petrol. Geologists Bull.*, **53**, 73–93.

6. CONNAN, J. (1974). Time–temperature relations in oil genesis, *Am. Assoc. Petrol. Geologists Bull.*, **58**, 2516–21.

7. DICKEY, P. A. (1975). Possible primary migration of oil from source rock in oil phase, *Am. Assoc. Petrol. Geologists Bull.*, **59**, 337–45.

8. DICKINSON, G. (1953). Geological aspects of abnormal reservoir pressures in Gulf Coast, Louisiana, *Am. Assoc. Petrol. Geologists Bull.*, **37**, 410–32.

9. DODSON, C. R. and STANDING, M. B. (1944). Pressure–volume–temperature and solubility relations for natural gas–water mixtures, *Drilling and Production Practice*, Am. Petrol. Inst., 173–8.

10. ENGELHARDT, W. V. and GAIDA, K. H. (1963). Concentration changes of pore solutions of clay sediments, *J. Sediment. Petrol.*, **33**, 919–30.

11. EVANS, C. R., MCIVOR, D. K. and MAGARA, K. (1975). Organic matter, compaction history and hydrocarbon occurrence—Mackenzie Delta, Canada, *Proc. 9th World Petroleum Congress*, Panel Discussion 3, 149–57.

12. FERTL, W. H. and CHILINGARIAN, G. V. (1976). Importance of abnormal formation pressures to the oil industry, S.P.E. 5946, *Soc. of Petrol. Engineers of AIME*.

13. GRATON, L. C. and FRASER, H. J. (1935). Systematic packing of spheres with particular relation to porosity and permeability, *J. Geology*, **43**, 785–909.

14. GUSSOW, W. C. (1962). Regional geological cross-sections of the western Canada sedimentary cover, Alberta Soc. of Petrol. Geologists, Calgary, Alta.

15. HEDBERG, W. H. (1967). Pore-water chlorinities of subsurface shales, Ph.D. thesis, Wisconsin Univ. (Univ. Microfilms, Ann Arbor, Michigan).

16. HOBSON, G. D. (1954). *Some Fundamentals of Petroleum Geology*. Oxford Univ. Press, London.

17. HUBBERT, M. K. (1953). Entrapment of petroleum under hydrodynamic conditions, *Am. Assoc. Petrol. Geologists Bull.*, **37**, 1954–2026.

18. HUNT, J. M. (1974). How deep can we find economic oil and gas accumulations?: S.P.E. 5177 in 1974, *Deep Drilling and Production Symposium Preprints*, 103–10.

19. JONES, P. H. (1967). Hydrology of Neogene deposits in the northern Gulf of Mexico basin, in: *Proc. 1st Symposium on Abnormal Subsurface Pressure*, Baton Rouge, Louisiana, Louisiana State Univ., 91–207.

20. KENNEDY, G. C. and HOLSER, W. T. (1966). Pressure–volume–temperature and phase relations of water and carbon dioxide, in: *Handbook of Physical Constants* (revised edn) Sect. 16, Geol. Soc. Am. Mem., **97**, 371–83.

21. LEVORSEN, A. I. (1967). *Geology of Petroleum*, 2nd edn, Freeman, San Francisco.

22. LEWIS, C. R. and ROSE, S. C. (1970). A theory relating high temperatures and over-pressures, *J. Petrol. Techn.*, **22**, 11–16.

23. Low, P. F. (1976). Viscosity of interlayer water in montmorillonite, *Soil Sci. Soc. of Am. J.*, **40**, 500–5.

24. Magara, K. (1968a). Compaction and migration of fluids in Miocene mudstone, Nagaoka Plain, Japan, *Am. Assoc. Petrol. Geologists Bull.*, **52**, 2466–2501.

25. Magara, K. (1968b). Subsurface fluid pressure profile, Nagaoka Plain, Japan, *Japan Petrol. Inst. Bull.*, **10**, 1–7.

26. Magara, K. (1971). Permeability considerations on generation of abnormal pressures, *Soc. of Petrol. Engineers J.*, **11**, 236–42.

27. Magara, K. (1972). Compaction and fluid migration in Cretaceous shales of Western Canada, *GSC Publication*, Paper 72–18.

28. Magara, K. (1974a). Compaction, ion-filtration, and osmosis in shales and their significance in primary migration: *Am. Assoc. Petrol. Geologists Bull.*, **58**, 283–90.

29. Magara, K. (1974b). Aqua-thermal fluid migration, *Am. Assoc. Petrol. Geologists Bull.*, **58**, 2513–16.

30. Magara, K. (1975a). Re-evaluation of montmorillonite dehydration as cause of abnormal pressure and hydrocarbon migration, *Am. Assoc. Petrol. Geologists Bull.*, **59**, 292–302.

31. Magara, K. (1975b). Importance of aqua-thermal pressuring effect in Gulf Coast, *Am. Assoc. Petrol. Geologists Bull.*, **59**, 2037–45.

32. Magara, K. (1976). Water expulsion from clastic sediments during compaction—directions and volumes, *Am. Assoc. Petrol. Geologists Bull.*, **60**, 543–53.

33. Martin, R. T. (1962). Adsorbed water on clay: a review, in: *Clays and Clay Minerals*, 9th Nat. Conf. Proc., Pergamon Press, New York, 28–70.

34. McKelvey, J. G. and Milne, I. H. (1962). The flow of salt solutions through compacted clay, in: *Clays and Clay Minerals*, 9th Nat. Conf. Proc., Pergamon Press, New York, 248–59.

35. Powers, M. C. (1967). Fluid-release mechanisms in compacting marine mudrocks and their importance in oil exploration, *Am. Assoc. Petrol. Geologists Bull.*, **51**, 1240–54.

36. Price, L. C. (1976). Aqueous solubility of petroleum as applied to its origin and primary migration, *Am. Assoc. Petrol. Geologists Bull.*, **60**, 213–44.

37. Salathiel, R. A. (1973). Oil recovery by surface film drainage in mixed-wettability rocks, *J. Petrol. Techn.*, **25**, 1216–24.

38. Schmidt, G. W. (1973). Interstitial water composition and geochemistry of deep Gulf Coast shales and sandstones, *Am. Assoc. Petrol. Geologists Bull.*, **57**, 321–37.

39. Skinner, B. J. (1966). Thermal expansion, in: *Handbook of Physical Constants* (revised edn) Sect. 6, Geol. Soc. Am. Mem., 97, 75–96.

40. Tissot, B. and Pelet, R. (1971). Nouvelles données sur les mécanismes de genèse et de migration du pétrole simulation mathématique et application à la prospection, *Proc. 8th World Petroleum Congress*, 35–46.

41. Van Olphen, H. (1963). Compaction of clay sediments in the range of molecular particle distances, in: *Clay and Clay Minerals*, 11th Nat. Conf. Proc.. Macmillan, New York, 178–87.

42. Vyshemirsky, V. S., Trofimuk, A. A., Kontorovich, A. E. and Neruchev, S.

G. (1973). Bitumoids fractionation in the process of migration, in: *Advances in Organic Geochemistry* (B. Tissot and F. Bienner, eds.), Technip., Paris, 359–65.

43. WILSON, H. H. (1975). Time of hydrocarbon expulsion, paradox for geologists and geochemists: *Am. Assoc. Petrol. Geologists Bull.*, **59**, 69–84.

Chapter 4

ESTIMATION OF THE MAXIMUM TEMPERATURES ATTAINED IN SEDIMENTARY ROCKS

BRIAN S. COOPER

Robertson Research International, Llandudno, UK

SUMMARY

Oil and gas are generated when sediments containing organic matter are heated to temperatures in excess of 75 °C during normal burial. The study of temperature gradients in presently subsiding sedimentary basins is useful in determining zones of hydrocarbon generation, and gives an insight into past temperature regimes in older basins. Particularly the effects of over-pressuring and fault-directed migrant hot connate waters are not to be overlooked. The actual determination of palaeo-temperatures can be made using vitrinite reflectivity, spore coloration and electron spin resonance of kerogen by correlating these indices with known temperature gradients, or by laboratory measurements of kinetic equations of change. However, there are possible sources of error both in measurement of indices and in fitting the indices to thermal models.

INTRODUCTION

Three basic parameters direct the course of the chemical reactions which change kerogen, that is sedimentary organic matter, into hydrocarbons. They are time, pressure and temperature. Biostratigraphy gives good control of the measurement of geological time intervals, and pressure, of least importance, can be readily measured or calculated. But temperature is a major and yet less well understood factor in defining the sequence of events which are initiated when a hydrocarbon source rock subsides, is buried and generates its hydrocarbons.

127

It is generally accepted that organic matter will begin to generate liquid and gaseous hydrocarbons between temperatures of 70 and 85 °C, with a transition to dry gas occurring between 150 and 175 °C, leading to the so-called 'liquid window' concept of Pusey.[1] The process of diagenetic change of organic matter is termed 'maturation', so that potential source rocks are immature if they have not been heated sufficiently to generate hydrocarbons, in the mature stage when they are releasing wet gas or oil, and post mature when they have passed through these stages and dry gas is the sole product of further diagenesis.

For the most part, the state of maturity is assessed by geochemical indices which relate to physical or chemical properties of the kerogen in the sedimentary sequence. However, it is becoming of increasing importance that these relative parameters be linked to temperature to understand the development of sedimentary basins and their history of hydrocarbon generation. The associated phenomena such as rift faulting, growth faulting, vulcanism, accelerated subsidence and over-pressure can all lead to changes in heat flow which will change the rate of maturation of kerogen. Since it is a prime aim of the petroleum geologist to relate the development of reservoir structures to the generation and migration of oil, it is not enough to know if the hydrocarbon source rocks are mature, but also when they became mature in relation to the reservoir structure and opening of migration paths. It is also important that the relationship between temperature and other phenomena such as over-pressuring and cementation should also be known. In order to carry out such determinations it is necessary to know the palaeo-temperatures which have been reached by sediments during their geological history and to express maturation in terms of time and temperature.

CONTEMPORARY SUBSURFACE TEMPERATURES

The study of present-day temperature regimes in and around sedimentary basins demonstrates its usefulness and *inter alia* some of the phenomena which must be borne in mind when reconstructing past thermal regimes. Most apposite to the search for oil, Klemme[2] has shown that many major hydrocarbon fields are associated with high geothermal gradients. It does not seem unreasonable to assume that hydrocarbons will be generated faster and be more likely to be captured in reservoirs if source rocks are undergoing a rapid increase in temperature by subsiding under the influence of a high geothermal gradient. But the causes of high geothermal gradients are numerous.

During the development of a rift valley system and its thermal bulge a generally higher geothermal flux would be expected centred on the rift valley and its volcanoes. However, apart from immediately adjacent to the volcanic areas, the heat flow is dissipated by deeply convecting waters which escape preferentially along active major faults, usually one of the rift boundary faults, and cause geothermal hot spots along their length. If rifting does not lead to plate separation, but to basin development as the thermal bulge contracts, these major faults can continue to be preferred upwards channels for fluids of deep origin. Alternatively, when rifts become zones of plate separation, the thermal bulge emerges as a mid-oceanic ridge with a strongly developed system of convecting waters within its flanks which cause localised hot spots where they reach the surface.[3] For a time these hot spots may underlie the earliest of the continental slope deposits.

The development of sedimentary prisms on the edges of continents, particularly if they are rapidly accumulating deltaic sediments, brings into play another combination of physical and tectonic features. The rapid burial of thick shale sequences, particularly if they contain montmorillonites and smectites, leads to under-compaction and over-pressuring so that fluid pressures approach lithostatic pressure, and to the development of growth faults.[4] A secondary effect, because of the low thermal conductivity of under-compacted and over-pressured shales, is the development of high geothermal gradients through them.[5] The growth faults become channels for the upward passage of fluids and cause localised hot spots as in Fig. 1, based on work in south Texas, from Jones and Wallace.[6] In a more detailed analysis of wells in a smaller area in south Louisiana, Kumar[7] concludes that over-pressuring causes fluids to break out along fault channels and then to be redistributed into permeable sediments at higher levels when further progress along the fault is impeded. Obviously the upwards distance through which fluids can move before being diverted depends upon the length of time a coherent column of water exists along the fault plane, and therefore on the permeability of both the providing and the receiving reservoir. In the example described geothermal highs occurred both over the fault planes and over adjacent structural highs.

Geothermal surveys in other oil provinces[2] have also shown interesting details, if only because of the questions left unanswered. In the North Sea area well data have been used to calculate geothermal gradients[8,9,10] and show that high values (35–50 °C/km) are associated with salt plugs and with areas of thick Tertiary sediments. Cornelius[10] points out that one thermal anomaly east of Yorkshire is associated with an earthquake centre. Both Harper[8] and Cornelius[10] conclude that the high geothermal gradients, which are based on bottom-hole temperatures, are caused by heat damming

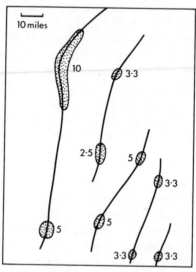

FIG. 1. Schematic sketch to show relation of hot spots to faults in south Texas; the numbers indicate in °F/100 ft the geothermal gradients in the shaded zones; background values range from 1·2 °F/100 ft to 2·0 °F/100 ft.

below under-compacted clays. A study of maximum palaeo-temperatures in the North Sea[11] suggests that high palaeo-temperature gradients are found in areas adjacent to fault zones and that temperatures had been reached, during the late Tertiary, higher than those now found. Although the data of Evans and Coleman[9] also show a general relation between Tertiary depocentres and high geothermal gradients, maximum values of geothermal gradient tend to lie near changes in strike of major structural faults.

The migration of fluids of deep origin and their thermal history are of particular significance in the exploration for ore deposits of the Mississippi Valley type. Such deposits tend to be found on the flanks of sedimentary basins. Deep connate waters, enriched in chlorides and metals, issue from the sediments during the late stages of compaction and make their way towards the basin edge, following deep faults upwards until at higher levels they deposit minerals by mixing with other fluids, by reacting with carbonates or by cooling. The minerals are likely to be from the suite pyrite, chalcopyrite, sphalerite, galena, fluorite and barite, and to be deposited at temperatures of between 50 and 250 °C. Both bitumens and oil inclusions are to be found with these deposits, and Dozy[12] has drawn attention to the parallelism in processes which produce both mineral deposits and hydrocarbons in the Mississippi Valley. To account for the repeated deposition of minerals in veins and the differences in temperature of the host rock and migrant fluids, Sibson et al.[13] have suggested that normal and transform fault planes can act as seismic pumps activated by earthquakes, and they draw attention to the episodic appearance of metal-bearing brines in the Red Sea.[14]

KEROGEN COMPONENTS

The most commonly used maturation indices are based on either the optical properties of particular kerogen components or on the chemical properties of the total kerogen. Kerogen is defined as that part of the organic matter of sediments not soluble in common organic solvents.[15] Usually only 5 to 20 % of the organic matter is soluble. Kerogen is also a mixture of types of organic matter of differing origins and contrasting chemical composition. The terminology of the coal petrologists is most often used to classify them and they are described in greater detail elsewhere,[16,17] but points of particular significance are briefly presented here.

Inertinites are charcoal-like substances with low oxygen and hydrogen contents. They show little change during maturation and have negligible hydrocarbon potential. Chemically they are almost graphite-like in structure, i.e. polycyclic aromatic, and are derived from land plant debris which has undergone prolonged microbial degeneration in soil. In appearance they are opaque, highly reflecting and optically isotropic.

Vitrinites are derived from land plant debris and humic acid gels which have accumulated in stagnant water. Telinite, representing remnants of cellular structure, and collinite representing solidified humic acid gels, are translucent yellow-orange in the early stages of diagenesis, and in incident light show low reflectivities (0·2 to 0·4 %). With increase in maturation, their colour turns to brown and opaque, and their reflectivities, and optical anisotropy develop. Vitrinites are significant gas sources and may generate

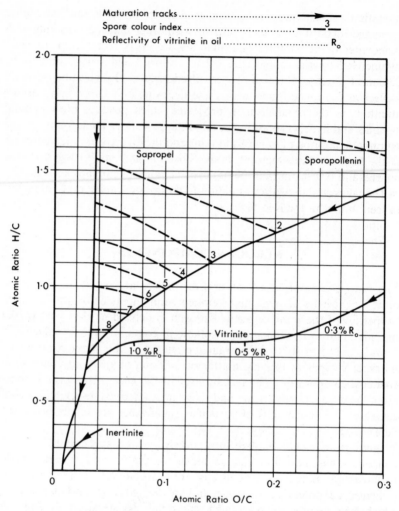

FIG. 2. Elemental ratios of main kerogen groups during maturation.

limited amounts of oil. Chemically they contain aromatic hydrocarbon structures linked by short aliphatic chains.

Exinites and liptinites comprise the remains of waxy and fatty plant tissues, and include sporinite, resinite, cutinite, alginite and amorphous sapropel. They are derived from spores and pollens, resin, plant cuticles and algal bodies, with amorphous sapropel being the disseminated and comminuted debris of algal and bacterial cells in sediment. After being extracted from rocks by dissolving the mineral components in acid, they have readily recognisable forms; amorphous sapropel appears as distinctive aggregates with slightly granular and waxy or oily texture. Chemically they are rich in aliphatic hydrocarbon groups joined in large molecular frameworks. During the early stages of diagenesis they are pale yellow in colour, but change through golden yellow, orange, brown to opaque, well illustrated by Raynaud and Robert.[18] They fluoresce similar colours in blue or UV light, and their reflectivities range upwards from 0·1% during the course of maturation. All of this group are potential sources of oil, with amorphous sapropel of most significance.

The chemical changes of each kerogen group are shown in Fig. 2, where it is seen that during the late stages of diagenesis they arrive at a common composition.

The kerogen in a sediment is composed of a mixture of these kerogen types, most of which will have been freshly deposited with the sediment, but some may have been reworked from older sediments.

PALAEO-TEMPERATURES FROM VITRINITE REFLECTIVITIES

Vitrinite in its various forms is the predominant component of coal, and it was coal petrologists who recognised that, through the coal rank series from peat to brown coal to bituminous coal to anthracite, vitrinites increased in brightness as seen on polished surfaces viewed in incident light. The introduction of stabilised light sources, sensitive photometers and high magnification to microscopes has enabled precise measurements to be made of the amount of light reflected from carefully polished plane surfaces of vitrinite.

During preparation the kerogen is concentrated using normal palynological techniques without oxidation, that is by dissolving the mineral components of the sediment in hydrochloric and hydrofluoric acids. The kerogen is mixed with cold-setting synthetic resin and spread either on a glass slide or into a shallow hole in a preformed resin block. Alternatively, if the relationship of the kerogen particles with their original mineral matrix is to be investigated, rock chips are immersed in the resin contained in a mould. After setting, the surfaces of the specimens are ground down on a plane surface using successively finer grades of carborundum and alumina.

Sections made on glass slides are particularly useful in that examination can be made in both transmitted and incident light, but it is advisable also to prepare a block from rock chips in order to see the petrographic relationships which may aid in the identification of unusual kerogen types.

For vitrinite reflectivity measurement, a high magnification photometric microscope with a stabilised quartz halogen incident light source is used, with accessories such as oil-immersion lenses, blue light fluorescence filters, or a UV light source and the ability to switch to transmitted light. Carefully polished glass standards of known reflectivity are used to standardise the apparatus, and the standards must themselves be checked fairly frequently against other standards, since some glasses are prone to surface attack by air.

In choosing particles to be measured, procedures vary. In some laboratories all organic particles are measured, and kerogen types are assigned to the distributions of reflectivities. Where skilled microscopists are available, it is usual to observe the general distribution of components and then, working sequentially over the surface of the sample, measure only vitrinite particles, making notes on their morphology and associations. Between twenty and fifty particles are measured. The results are plotted on histograms and, for example, from a drill cuttings sample would be likely to show a dominance of indigenous vitrinite, with lesser amounts of lower reflecting vitrinite from cavings and higher reflecting vitrinite from reworked particles.

The attractiveness of the vitrinite reflectivity method lies in its ability to give precise numerical measurements, although, as has been seen, some interpretative judgement by the microscopist may have already been made. Subjective judgement is also called for in the assessment of the vitrinite gradient for a well section so that the presence of coals, well defined horizons and casing points are useful reference points. However, the behaviour and constitution of vitrinite or vitrinite-like materials may vary between adjacent lithologies. The vitrinite of the Carboniferous coals of north-eastern England have substantially lower reflectivities than coals of similar age and rank in Germany, possibly due to an impregnation at molecular level of resinite. In bituminous and hot shales, fossil wood or jet is not unusual and has the appearance of vitrinite, but with much lower reflectivity than that of adjacent coals. These shales often contain collinite-like laminae, but again with low reflectivities, and it is conjectured that both these and jet are impregnations of bitumen-like material.[19] The reflectivity of vitrinite also depends upon the nature of the enclosing sediment,[20] and usually dispersed vitrinite particles in shales, sandstones and limestones have lower reflectivities than the vitrinites of associated coal seams.

The calculation of palaeo-temperatures from coal rank originated with the laboratory experiments of Huck and Karweil,[21] with amplifications by Karweil,[22] Teichmuller[23] and Bostick.[24] Other estimates have been made by Lopatin,[25] Hood et al.[26] and others. The nomogram (Fig. 3) illustrates the relationship between time, temperature and vitrinite reflectivity, and shows how heating at a constant temperature for a given time produces a particular value of reflectivity. Under normal conditions of basin development, however, the temperature of a subsiding horizon will increase at varying rates dependent upon the rate of subsidence and changes in thermal flux and thermal conductivity. To calculate these effects one can assume that the temperature increase takes place in steps, separated by finite periods of constant temperature. Examples of such calculations are given in Doebl[27] and Karweil.[28] Bostick,[24] on the other hand, suggests that the reflectivity change due to increase in temperature from T to $T + \Delta T$ in time t is the same as if the vitrinite was heated at constant temperature $T + 0.65\Delta T$ for time t. However, any calculation of step by step maturation shows that, in most cases, the period of time at which the highest range of temperatures have acted on the vitrinite is the most critical. Hood et al.[26] have taken the time when the vitrinite is within 15% of the maximum temperature as the most useful parameter and developed a scale, 'Level of Organic Maturity' or LOM, which with values between 0 and 20 gives a linear increase with depth for sediments undergoing burial at constant rates of subsidence and at a constant temperature gradient. They show that there is a linear-logarithmic relationship for maximum temperature reached and effective time of heating for any value of LOM. Because of assumptions concerning the values of activation energies, conclusions derived from LOM values, which are directly correlated with vitrinite reflectivity, are only accurate for LOM values between 9 and 16.

To derive a nomogram which relates rank change to rate of heating is not impossible, since one can describe the isothermal changes in a mathematical form so long as it is susceptible to differential calculus, although the mathematics may not have any connection with the reaction kinetics. An example is given in Fig. 4, based on Karweil's diagram. A change in temperature plotted on the right hand side from T_1 to T_2 allows a power function P to operate; if P is operative for t million years, the product Pt will cause a vitrinite originally of R_1 reflectivity to increase to R_2. This method, like other methods, has to be checked against experience, but such tools are of use when their limitations are understood.

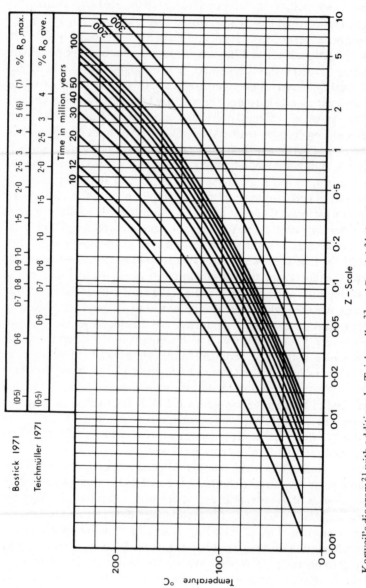

FIG. 3. Karweil's diagram[21] with additions by Teichmuller[23] and Bostick[24] showing vitrinite reflectivities produced by heating at constant temperatures.

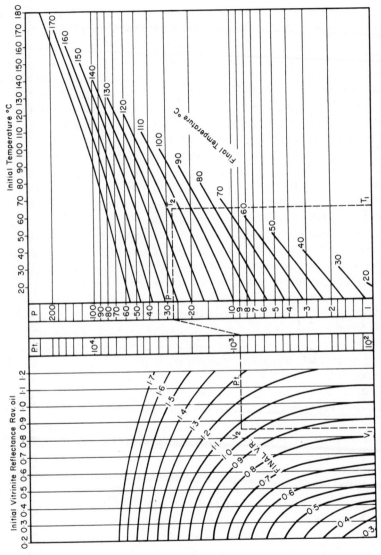

FIG. 4.   Nomogram relating change in vitrinite reflectivity in oil to changes in temperature and time.

SPORE COLORATION

Early attempts at evaluating the maturity of organic matter was by observing the colour of isolated kerogen. Since, however, kerogen components are of various colours, the method was somewhat imprecise. In order to obtain stratigraphic control in ditch cuttings and to compare similar forms, measurements of colour are now carried out by palynologists on spores and pollens only. Thus, anomalies due to caving and reworking are resolved. The techniques pioneered by Gutjahr,[29] and developed by Correia[30] and Staplin,[31] are widely used. Preparation is by normal palynological methods of digestion in hydrochloric and hydrofluoric acids, but without the use of oxidising acids. After centrifuging in zinc bromide the kerogen is made into a normal slide mount and observed in transmitted white light. Either photometric measurements of transmitted colour passing through green, red or blue filters are made, or there is direct comparison with standard mounts.

With increasing maturation, spore colours change from pale yellow, through golden yellow, orange, brown to opaque and it is possible to distinguish 12 or so stages of colour development. The method has been criticised as subjective, and standards of colour appreciation may vary from laboratory to laboratory, but it is found that with common or similar standards and a regular interchange of material, groups of operators can achieve close reproducibility. As a technique for estimating the maturity of oil-generating kerogen, the chief advantage is that sporopollenin is chemically closer to sapropel than vitrinite, and may give a true indication of the stage of hydrocarbon generation. Indices of measurement vary between authors; both Staplin and Correia have used a 0 to 5 scale based on distinctive changes in colour, but Barnard et al.[32] have published a ten-point scale based on an average linear increase with depth.

Interpretation of spore colour indices usually presents no problems, but anomalies do occur. In red beds, palynomorphs may have been oxidatively bleached during deposition and retained a paleness of colour during maturation. Strongly coloured palynomorphs are seen in bituminous and 'hot' shales, due probably to the uptake of heavy metals. These may also occur in reservoir rocks; reservoir rocks often show a wider range of colours than shales, possibly due to the intermittent passage of heated migrant fluids. Passing through overpressured zones, spore colour indices often increase rapidly, although vitrinite reflectivity is less prone to such rapid change, and suggests that sporopollenin is more susceptible to actual temperatures than to the duration of heating.

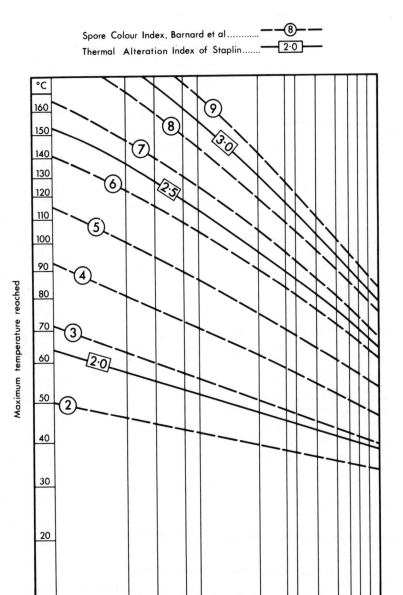

Spore Colour Index, Barnard et al............ — — ⑧ — —
Thermal Alteration Index of Staplin....... [2·0]

FIG. 5. Spore colour changes with increasing temperature; spore colour index of Barnard *et al.*,[32] thermal maturation index of Staplin.[31]

It has been possible to correlate spore colours with maximum palaeo-temperatures by using bottom-hole temperatures and vitrinite-derived temperatures (Fig. 5), although there is no theoretical work for sporopollenin like that of Karweil for vitrinite. The figure appears to show a much greater dependence on temperature and lesser dependence on time of heating than vitrinite.

ELECTRON SPIN RESONANCE MEASUREMENTS ON KEROGEN

The aromatic hydrocarbon structures in kerogen carry free electrons stabilised by the resonating bonds of benzene rings. The number and distribution of free electrons in kerogens depends on the number and relative position of the benzene rings. If a sample of kerogen is placed in the path of a microwave and under the influence of a magnetic field which can be varied, at particular magnetic field strengths the free electrons will resonate and alter the microwave frequency. Using an electrometer, the number of free electrons per gramme, Ng, the location of the resonance point, g, and the width of the signal, w, can be determined.

The first studies on electron spin resonance of kerogens were by Marchand et al.,[33,34] which showed progressive changes in these measured parameters during the course of maturation. Using measured temperatures in wells of currently subsiding basins, Pusey[1] was able to correlate Ng and g with maximum temperature and extend the method to older sediments. The method is based on the changes in chemical composition of kerogen during maturation in which aromatic hydrocarbon structures become more frequent and larger in size, and assumes that freshly deposited organic matter will be low in benzene groups. Durand et al.[35] have pointed out some of the limitations of the method, and particularly that the composition of the kerogen may be critical. Humic acids and humified lignin contain significant amounts of both benzene groups and oxygen, and are likely to give non-consistent Ng and g values. The inertinites can give very strong ESR signals, as can reworked organic matter. Nevertheless, the method has been used with success in the Gulf of Mexico, south-east Asia and many other areas where normal marine sediments contain kerogens which are dominantly sapropelic and vitrinitic. The method uses very little material. The kerogens—isolated in cold acids—do not have to be very pure, and the method is fast.

In a study of Mesozoic and Tertiary sediments of the Gulf Coast area Pusey[36] was able to show maximum palaeo-temperatures were up to 50 °F higher than present temperatures and could be accounted for by erosion after uplift or by higher geothermal gradients in the past. A study of wells in the northern North Sea by this method[11] also indicated past, higher geothermal gradients, particularly in the vicinity of fault zones.

OTHER METHODS

A technique used by ore mineralogists which could be attractive to petroleum geologists, is the determination of palaeo-temperatures by fluid inclusion thermometry.[37] Fluid inclusions occur in gangue minerals such as quartz and calcite, which also may be deposited as cements in reservoir rocks. The fluid usually partially fills a near-spherical cavity in the mineral grain, and by using a freezing and hot stage microscope both the salinity of the aqueous fluid and its homogenisation temperature can be determined. The homogenisation temperature, that is the temperature at which the fluid and vapour phases coalesce, is the minimum temperature of deposition of the mineral. If the depth of overburden can be roughly estimated, perhaps by age-dating associated clays or by inference from the fluid inclusion salinity, an accurate palaeo-temperature can be calculated using a pressure correction.[38] The correction is not usually very large. To be able to relate the cementation of reservoirs to other events would be quite useful.

Currie and Nwachukwu[39] investigated by this method the minimum temperature of deposition of quartz in fractures in the Cardium sandstone of well sections in western Canada. By assuming that the fractures were opened soon after the sandstone had reached its maximum palaeo-temperature, and using shale transit times to calculate maximum overburden and palaeo-pore pressures, Magara[40] has been able to estimate the actual value of the maximum palaeo-temperatures.

The ongoing interest in montmorillonite and smectites, and their conversion towards illite, is beginning to give hopes that another palaeo-temperature indicator will emerge. It appears that montmorillonite changes to montmorillonite–illite at between 80 and 120 °C,[41] and that the transition is dependent on pressure. Again it would be useful to have information on the temperature–mineral composition relations and be able to fit montmorillonite–smectite generated over-pressures into the geothermal history of basin subsidence and tectonism.

DISCUSSION

It has been usual for geochemists to assume that all kerogen types behave similarly under the influence of time and temperature, so that, for example, it is frequently implied that sapropel will begin to generate oil when associated vitrinite has reached a reflectivity of 0·5 %. However, the chemical structures of vitrinite, sporopollenin and sapropel are different enough to suggest that they would depart from parallel paths of maturation and in Fig. 6 examples are given of vitrinite reflectivities plotted against spore colour indices for basins with contrasting depositional histories. Tissot and Espitalié[42] have attacked this problem most energetically and have produced a mathematical model of maturation in which it is assumed that each kerogen type is made up of different proportions of smaller reactive units. By applying the Arrhenius equation to each unit, its reaction rate can be described in terms of its activation energy and its Arrhenius factor. The activation energies, E, range between 10 and 80 kcal/mole, and the Arrhenius factors, A, between 10^4 and 10^{35} per million years. By using such a model it is possible to calculate the quantity of oil and gas formed by

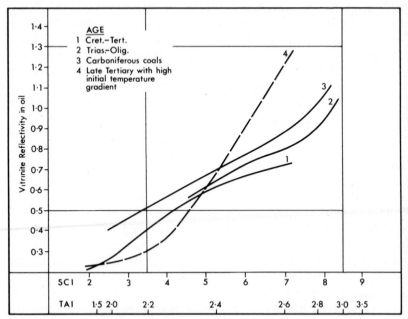

FIG. 6.   Vitrinite reflectivities plotted against spore colour indices.

source rocks in all parts of a sedimentary basin as a function of temperature, and to calculate the previous geothermal energy. Significantly, the model also demonstrates that dominant reactive units of the major kerogen types are different in their chemical potential as shown below.

	$\log A$	E kcal/mole
Vitrinite	32	60
Sporopollenin	27	50
Sapropel	31	70

To what extent the various maturation indices and the palaeotemperatures which they indicate depart from the actual temperatures at which oil is generated, remains to be discussed by geologists and geochemists. Certainly geothermometry will be used with greater facility in sub-soil exploration and in detecting migrant fluids, and palaeogeothermometry will play an important part in unravelling the sequence of events which lead to the generation of hydrocarbons and their emplacement in reservoirs.

ACKNOWLEDGEMENTS

I wish to thank the directors of Robertson Research International for permission to publish this communication.

REFERENCES

1. PUSEY, W. C. III. (1973). The ESR-kerogen method. How to evaluate potential gas and oil source rocks, *World Oil*, **176**, 71.
2. KLEMME, H. D. (1975). Geothermal gradients, heat flow, and hydrocarbon recovery, in: *Petroleum and Global Tectonics* (A. G. Fischer and S. Judson, eds.), Princeton Univ. Press, Princeton, New Jersey.
3. MORGAN, W. J. (1975). Heat flow and vertical movements of the crust, in: *Petroleum and Global Tectonics* (A. G. Fischer and S. Judson, eds.), Princeton Univ. Press, Princeton, New Jersey.
4. CURTIS, D. M. (1970). Miocene deltaic sedimentation, in: *Deltaic Sedimentation, Modern and Ancient*, Soc. Econ. Palaeontologists Mineralogists Spec. Pub., **15**, Tulsa.
5. LEWIS, C. R. and ROSE, S. C. (1970). A theory relating high temperatures and over-pressures, *J. Petrol. Tech.*, **22**, 11.

6. JONES, P. H. and WALLACE, R. H. (1974). Hydrogeologic aspects of structural deformation in the northern Gulf of Mexico basin, *J. Research US Geol. Survey*, **2**, 511.

7. KUMAR, M. B. (1977). Geothermal and geopressure patterns of Bayou Carlin—Lake Sand area, South Carolina: Implications, *Am. Assoc. Petrol. Geologists Bull.*, **61**, 65.

8. HARPER, M. L. (1971). Approximate geothermal gradients in the North Sea basin, *Nature* (London), **230**, 235.

9. EVANS, T. R. and COLEMAN, N. C. (1974). North Sea geothermal gradients, *Nature* (London), **247**, 20.

10. CORNELIUS, C-D. (1975). Geothermal aspects of hydrocarbon exploration in the North Sea area, *Norge Geologiske Undersokelse*, **316**, 29.

11. COOPER, B. S., COLEMAN, S. H., BARNARD, P. C. and BUTTERWORTH, J. S. (1975). Palaeo-temperatures in the northern North Sea basin, in: *Petroleum and the Continental Shelf of NW Europe* (A. W. Woodland, ed.) Applied Science Publishers, Barking, England.

12. DOZY, J. J. (1970). A geological model for the genesis of the lead–zinc ores of the Mississippi Valley, USA, *Trans. Inst. Min. Metall. (Sect. B: Appl. Earth Sci.)*, **7**, B163.

13. SIBSON, R. H., MOORE, J. McM. and RANKIN, A. H. (1975). Seismic pumping—a hydrothermal fluid transport mechanism, *J. Geol. Soc. Lond.*, **131**, 653.

14. BIGNELL, R. D. (1975). Timing, distribution and origin of submarine mineralisation in the Red Sea, *Trans. Inst. Min. Metall. (Sect. B: Appl. Earth Sci.)*, **84**, B1.

15. FORSMAN, J. P. and HUNT, J. M. (1958). Insoluble organic matter (kerogen) in sedimentary rocks, *Geochim. Cosmochim. Acta*, **15**, 170.

16. STACH, E. (1975). *Stach's Textbook of Coal Petrology*, 2nd edn, Gebruder Borntraeger, Berlin and Stuttgart.

17. COOPER, B. S. and MURCHISON, D. G. (1969). Organic geochemistry of coal, in: *Organic Geochemistry, Methods and Results* (G. Eglinton and M. T. Murphy, eds.), Springer-Verlag, Berlin.

18. RAYNAUD, J-F. and ROBERT, P. (1976). Les methodes d'etude optique de la matiere organique, *Bull. Centre Rech. Pau-SNPA*, **10**, 109.

19. TEICHMULLER, M. (1974). Origin and alteration of bituminous matter in coals and the relationship to origin and maturation of petroleum, *Fortschr. Geol. Rheinland Westfalen*, **24**, 65.

20. BOSTICK, N. H. and FOSTER, J. N. (1975). Comparison of vitrinite reflectance in coal seams and in kerogen of sandstones, shales and limestones in the same part of a sedimentary section, in: *Petrographie de la Matiere Organique des Sediments, Relations avec la Palaeo-temperature et le Potential Petrolier*, CNRS, Paris.

21. HUCK, G. and KARWEIL, J. (1955). Physikalisch-chemische probleme der inkohlung, *Brennstoff. Chem.*, **36**, 1.

22. KARWEIL, J. Z. (1955). Die metamorphose der Kohlen vom Standpunkt der physikalischen chemie, *Dtsch. Geol. Ges.*, **107**, 132.

23. TEICHMULLER, M. (1971). Anwendung kohlenpetrographischer methoden bei der erdöl- und erdgasprospecktion, *Erdol Kohle/Brennstoff-Chemie*, **24**, 69.

24. BOSTICK, N. H. (1971). Thermal alteration of clastic organic particles as an indicator of contact and burial metamorphism in sedimentary rocks, *Geosci. Man.*, **3**, 83.
25. LOPATIN, N. V. and BOSTICK, N. H. (1974). The geological factors in coal catagenesis, *Illinois Geol. Survey Reprint*, 1974-Q.
26. HOOD, A., GUTJAHR, C. C. M. and HEACOCK, R. L. (1975). Organic metamorphism and the generation of petroleum, *Am. Assoc. Petrol. Geologists Bull.*, **59**, 986.
27. DOEBL, F., HELING, D., HOMANN, W., KARWEIL, J., TEICHMULLER, M. and WELTE, D. (1975). The diagenesis of Tertiary clay sediments and included dispersed organic matter in relation to geothermics of the upper Rhine graben, in: *Approaches to Taphrogenesis*, Interunion Commission on Geodynamics Sci. Rep. No. 8, Stuttgart.
28. KARWEIL, J. (1975). The determination of palaeo-temperatures from the optical reflectance of coaly particles in sediments, in: *Petrographie de la Matiere Organique des Sediments, Relations avec Palaeo-temperature et le Potential Petrolier*, CNRS, Paris.
29. GUTJAHR, C. C. M. (1966). Carbonisation measurements of pollen grains and spores and their application, *Leidse Geol. Meded*, **38**, 1.
30. CORREIA, M. (1971). Diagenesis of sporopollenin and other comparable organic substances: application to hydrocarbon research, in: *Sporopollenin* (J. Brooks, P. R. Grant, H. Muir, P. van Gijzel, and G. Shaw, eds.) Academic Press, London.
31. STAPLIN, F. L. (1969). Sedimentary organic matter, organic metamorphism, and oil and gas occurrence, *Bull. Can. Petrol. Geol.*, **17**, 47.
32. BARNARD, P. C., COOPER, B. S. and FISHER, M. J. (In press). Organic maturation and hydrocarbon generation in the Mesozoic sediments of the Sverdrup Basin, Arctic Canada, *Proc. 4th Int. Palynol. Congress*, Lucknow.
33. MARCHAND, A., LIBERT, P. A. and COMBAZ, A. (1968). Sur quelques criteres physico-chimique de la diagenesis d'un kerogene, *Compt. Rend. Ser. D.*, **266**, 2316.
34. MARCHAND, A., LIBERT, P. A. and COMBAZ, A. (1969). Essai de caracterisation physico-chimique de la diagenesis de quelques roches organiques, biologiquement homogenes, *Rev. Inst. Petrol.*, **24**, 3.
35. DURAND, B., MARCHAND, A., AMIELL, J. and COMBAZ, A. (1976). Etude de kerogenes par RPE, in: *Advances in Organic Geochemistry*, 1975, Pergamon Press, London.
36. PUSEY, W. C. III. (1973). Palaeo-temperatures in the Gulf Coast using the ESR –kerogen method, *Gulf Coast Ass. Geol. Soc. Trans.*, **23**, 195.
37. ROEDDER, E. (1967). Fluid inclusions as samples of ore deposits, in: *Geochemistry of Hydrothermal Deposits* (H. L. Barnes, ed.), Holt, Rinehart and Winston Inc., New York.
38. OHMOTO, H. and RYE, R. O. (1970). The Bluebell Mine, British Columbia, 1. Mineralogy, paragenesis, fluid inclusions and the isotopes of hydrocarbon, oxygen and carbon, *Econ. Geol.*, **65**, 417.
39. CURRIE, J. B. and NWACHUKWU, S. O. (1974). Evidence of incipient fracture porosity in reservoir rocks at depth, *Bull. Can. Petrol. Geol.*, **22**, 42.
40. MAGARA, K. (1976). Thickness of removed sedimentary rocks, palaeo-pore

pressure and palaeo-temperature, south-western part of western Canada Basin, *Am. Assoc. Petrol. Geologists Bull.*, **60**, 554.

41. BURST, J. F. (1969). Diagenesis of Gulf Coast clayey sediments and its possible relation to petroleum migration, *Am. Assoc. Petrol. Geologists Bull.*, **53**, 73.

42. TISSOT, B. and ESPITALIÉ, J. (1975). L'evolution thermique de la matiere organique des sediments: applications d'une simulation mathematique, *Rev. Inst. Français Petrol.*, **30**, 743.

Chapter 5

SANDSTONES AS RESERVOIR ROCKS

J. C. M. Taylor

V. C. Illing & Partners, Surrey, UK

SUMMARY

This chapter reviews developments bearing on the prediction, delineation, and qualitative assessment of sandstone reservoirs that have taken place mainly in the past decade in the study of sedimentary environments, of diagenesis, and also in such diverse fields as seismic exploration, wireline logging and plate tectonics. Many of the advances are themselves a result of offshore exploration and deeper drilling.

The relation between depositional environments and reservoir quality is investigated. Attention is concentrated on sedimentary attributes that are likely to be recognisable in the subsurface. Migration of many types of sand body destroys the original geometry though preserving internal structures and fabrics relevant to their performance as reservoirs.

Scanty but growing evidence links the character of early diagenesis with specific depositional environments. Predicting the modification of reservoir properties by subsequent deeper burial can be approached by statistical methods, by an understanding of the physics and chemistry of the changes taking place, or by deductions from quantitative petrography, each method having limitations.

INTRODUCTION

Despite the outstanding performance of many carbonate reservoirs, sandstones remain the most important hosts to hydrocarbon accumulations. Over 60 % of the reserves in the world's giant fields (that is,

147

those containing more than 500 million barrels recoverable oil or its equivalent in natural gas, and which account for 65–70% of present production) occur in sandstone reservoirs; if we exclude the Middle East with its vast limestone fields, the proportion is 80%.[65]

It is not difficult to see why this should be, bearing in mind that it is not merely the existence of hydrocarbon accumulations that counts, but our ability to discover them. Initially-porous siliciclastic sands are formed in a broader range of environments than carbonates, which require specialised conditions to be formed at all. They are composed of relatively stable components, less subject than carbonates (although by no means immune) to physical and chemical changes leading to a loss of porosity once they are deposited. Given the tectonic history of a region in the kind of detail revealed by reconnaissance geology and geophysics, it is easier to indicate which areas are likely to have yielded arenaceous sediment and where it will have tended to accumulate, than to predict the distribution of porous carbonates. Favoured sites for the preservation of silicilastic sands include those types of rapidly subsiding depressions which tend to favour contiguous deposition of potential petroleum source rocks together with conditions favourable for the generation and release of petroleum from them.

In the past decade the industry has shown increasing awareness of the importance of the more subtle geological factors which influence the disposition, size and quality of sandstone reservoirs. Motivation has been provided by the ever-increasing demand for reserves, particularly through the shift from onshore to offshore exploration, and as a result of greater drilling depths. It has been facilitated by new theoretical concepts— developed both within the industry itself and in the academic world—by improved techniques and instrumentation, and not least by the influence of a fresh generation of petroleum geologists trained to accept sedimentology as an exploration tool, and incidentally the first to have been exposed as students to the paradigm of Plate Tectonics.

Offshore exploration has had several interlocking effects. Drilling and production costs are enormous, providing a strong incentive to locate reservoirs of high productive potential and drain them with the fewest wells possible. At the same time, particularly in the grabens associated with Atlantic-type ocean margins like those of the North Sea, subsurface geology often bears little or no resemblance to that on adjacent land areas, so that in the early stages of exploration the existence of potential reservoir horizons has to be predicted largely by using theoretical sedimentation models, guided by geophysics. Fortunately, seismic exploration at sea is cheaper and more effective than on land and can more often provide useful lithological information and indicate conditions favourable to stratigraphic as well as structural trapping. Deeper drilling, whether on land or at sea, as well as increasing the economic incentive to efficient exploration, has also focused attention on the effects of deep burial on reservoir porosity and permeability.

Sand bodies are responsible for a variety of types of stratigraphic trap. It is claimed with undeniable logic by their devotees[46] that more petroleum may be preserved in stratigraphic than in structural traps. As the number of large untested structures dwindles, it is to be hoped that this prediction is right. However, managements by and large remain unimpressed. Onshore, it is difficult to get approval for drilling a prospect that is not surrounded by concentric depth contours; offshore, it is virtually impossible. So long as serendipity continues to be the major factor in the discovery of stratigraphic traps[64] it has to be admitted that this attitude is defensible. Undoubtedly the application of sedimentology to exploration thinking should enhance the chances of intentional discovery, yet the magnitude of any improvement may be more dependent on developments in reflection seismic techniques leading to the resolution of finer detail, than on sophisticated sedimentology.

Consequently the present chapter will not be concerned primarily with stratigraphic traps as such, which have been extensively covered elsewhere,[64] but rather with the application of advances in sedimentology to the wider aspects of petrophysical properties on the one hand and reservoir morphology on the other, applicable to both stratigraphic and structural traps.

SPECIAL CHARACTERISTICS OF SANDSTONES AS RESERVOIR ROCKS

The particulate nature of sand and the manner in which it is distributed impose distinctive features on sandstone reservoirs, which consequently often differ markedly in macroscopic and microscopic geometry and in physico-chemical behaviour from carbonate reservoirs. Thus, a characteristic of most sandstone reservoirs is an emphasis on lateral rather than vertical dimensions. In typical oilfield examples in marine sequences, individual sand units are seldom as much as 100 m and commonly less than 40 m thick, yet even narrow belts rarely measure less than 2 km across. It is true that uninterrupted sands several hundreds of metres thick are known, but they are generally of continental origin and so are rarely favourably situated for hydrocarbon accumulation; when, however, other circumstances are propitious, very large and productive reservoirs that are economical to develop may result.

The microscopic geometry of sands sets limits to porosity, permeability and hydrocarbon saturation. The pores through which all—or almost all—fluid movement must occur are the obverse of packings of sub-spherical particles up to about 2 mm diameter.

The properties of such packs have been investigated from a mathematical standpoint, experimentally, and in natural sands by a number of workers. Rigorous treatment is difficult because natural sands do not pack in the closest, most stable, nor indeed any regular manner. Even in the case of spheres Graton and Fraser[44] found that natural packings are usually colonies of rhombohedral packing (the most compact arrangement possible, with a porosity of 26%) in a surrounding mesh of haphazard packing. The geometry of prolate spheroids, which approximate more closely than spheres to real sand grains, has since been examined by Allen.[2]

In regular packing arrangements porosity is independent of grain size, but the type of packing achieved in real sands is dependent on shape, roundness, and sorting of the grains—properties which are often correlated with grain size. Beard and Weyl[7] have investigated the effects of grain size and sorting on porosity and permeability of artificially mixed and packed river sands and found that the porosity was essentially independent of grain size for wet-packed sand of the same sorting, but fell with poorer sorting from about 42% for extremely well sorted sand to 28% for very poorly sorted sand. They considered that the probable effect of low sphericity and high angularity is to increase porosity and permeability.

Most of the experimental work published confirms that the porosities of sands at or shortly after deposition fall in the range 30–50% and cluster strongly round a value of 40%. Petrographic evidence from many thin sections, in which early cementation has retained the original packing, suggests that this was also true of ancient sands in a wide range of sedimentary environments. The packing density—originally defined in a different but statistically equivalent form by Kahn[62]—can be considered to be the complement of the porosity in a clean cement-free sand. Thus, an initial packing density of approximately 60% can be used as a rough yardstick to judge whether the grains in a sandstone were likely to have been separated by matrix at the time of deposition, or conversely, whether significant compaction has since occurred. Most reservoir sandstones have porosities appreciably lower than 40 %, mainly as a result of cementation rather than compaction. Values of between 20 % and 30 % are common in economic North Sea fields.

Permeability is a function of grain size and sorting. According to Krumbein and Monk:[66]

$$K = Cd^2 \exp(-1 \cdot 35\sigma)$$

where K is the permeability, C a constant, d the geometric mean grain diameter and σ the standard deviation of the grain size about the mean. Beard and Weyl,[7] in their experiments on packed sands, found that permeabilities increased by a factor of four for each increase in Wentworth grade size, in agreement with Krumbein and Monk's results. The maximum average measured value was 475 darcys for an extremely well sorted coarse sand, but only 14 darcys for a very poorly sorted coarse sand. Permeabilities in sandstone reservoirs are significantly lower; anything over 1 darcy is considered good, and values over 4 darcys are exceptional. The discrepancy is partly due to cementation and partly to departures from the requirement for d'Arcy's law that there should be no reaction between fluid and solid media. Sandstones commonly consist predominantly of mildly hydrophilic substances (quartz, carbonates), together with variable amounts of strongly hydrophilic material (clay minerals). Clays also have a mechanical effect on permeability out of proportion to their volume, as a result of their platy or fibrous particle shape. In deep reservoirs compressibility of the pore system also needs to be taken into account.[31]

The constituent grains of a sandstone are usually slightly elongated, and a number of studies have shown that the longer axes tend to be aligned parallel with the depositing current direction. Mast and Potter[74] have shown that this leads to a slightly increased permeability in the direction of grain elongation.

With some well-documented exceptions sandstone reservoirs are water-wet; the water being held as a thin film around grains, and in greater volumes in collars at grain contacts and in pores bounded by smaller than usual throats.[54] Minimum water saturations of producing sands are seldom less than 10 %, and commonly fall in the range 15–40 %. Small amounts of clay can greatly increase the amount of water held in the pores, and according to Pirson[83] shaly sands can produce water-free oil when their water saturation is as high as 65 %. Small-scale packing heterogeneities are probably the chief factor responsible for the levels of irreducible water saturation in clean sands; their effects are described by Morrow.[79]

Paradoxically, the higher the porosity and permeability of a reservoir sandstone (i.e. the less cemented it is) the less likely is it that fractures or joints will create paths of higher fluid conductivity; in this sense, therefore, porosity and permeability in sandstone reservoirs tend to be more homogeneous than in carbonate reservoirs where fissures and fractures may provide virtually hydrostatic communication between wells, kilometres apart, despite low matrix porosities and permeabilities. Inhomogeneity is often important in sandstones too, but takes the form of local barriers in a more porous and permeable medium rather than local channels of communication in a tighter one.

DEPOSITIONAL FACTORS CONTROLLING SANDSTONE RESERVOIR GEOMETRY AND PROPERTIES

The thickness of sand bodies, their lateral extent, their shape in plan, internal structure, and the nature of their boundaries against enclosing strata, are partly determined by the depositional environment, which thus exerts an influence on reservoir volume and the positioning of wells. Great strides have been made in the past ten years in the interpretation of sedimentary sequences in terms of a limited number of environmental or facies models.[96,102,119] These can be used, when one or more wells have been drilled, to predict the likely extensions and changes in character in different directions of any sand bodies encountered, and sometimes to locate bodies which have been missed.

In a sense it is scarcely possible to have too much geological knowledge about a prospect, particularly when secondary or tertiary recovery methods are contemplated. Yet there are dangers in attempting to apply too closely to subsurface problems models based on present sedimentation and outcrop studies. One is the impossibility of acquiring subsurface information on a scale commensurate with the size of the model; initial decisions on the development of a field covering an area 4 km by 8 km might have to be made on the basis of three holes yielding cores of about 10 cm across—a sampling density of 1 in 1.2×10^9! A warning against over-prediction—the insertion of details unwarranted either by the data or by the inferred process—is given by Moore.[77] These misleading details may arise from unique initial conditions or the interaction of other unrecognised processes. Moore notes, however, that when local variations are eliminated, the remaining message appears at first disappointingly simple

—but that experience shows these concepts have a power out of all proportion to their degree of elaboration.

These considerations apply with particular force to predictions of the *geometry* of sand bodies. When looking for analogues in Recent sands, it needs to be constantly borne in mind that the shape of a sand body in the geological record is the locus of its changing shape with time, modified by compaction, and perhaps later by tectonics and erosion. Except, perhaps, in deep water most sand bodies migrate, and in so doing change both their original geometry and the proportions of the depositional facies preserved; in extreme but very important cases this migration leads to the formation of sheet sands in which only a limiting thickness, together with a characteristic vertical succession, survive from the depositional geometry. At the other— and equally important—extreme, where subsidence relative to sea level matches the rate of accumulation, facies belts remain almost stationary and sand bodies grow vertically whilst retaining roughly their depositional length and breadth. Thus, the primary three-dimensional shape and size of sand bodies is of direct oilfield importance only in those rare instances when it is frozen in an instant of geological time, usually by rapid subsidence: such accidents, especially those involving entombment in marine shales, are of course important as potential stratigraphic traps.

Such qualifications apply mainly to the gross geometry of sandstone bodies. When due allowance is made for preservation potential, however, knowledge of the specific environment may be very helpful in understanding other factors of importance to reservoir quality such as internal stratification, texture, porosity and permeability, and the uniformity of these characteristics. It will be the thesis of a later section of this chapter that these distinctions are by no means rendered irrelevant by the influence of later diagenesis.

Before embarking on a discussion of the influence of the depositional environment on sandstones as reservoirs, it would be as well to remind ourselves what properties are important for optimum oil extraction. Large reservoir size is obviously desirable, since it governs total reserves. High porosity and low minimum water saturation are important for the same reason. High rates of production are economically as important as volume of reserves, and although high permeability is therefore desirable, relatively poor permeability can be compensated by a sufficiently thick productive section.

Finally, the continuity and regularity of the pore system has to be considered. Arguably the best arrangement would be an isotropic reservoir in which the pores were all uniform in size and shape in all directions. Not only is permeability then a maximum for a given grain size and packing, but its uniformity reduces the chance that oil pockets will be overtaken by advancing edge or bottom water during production; consequently the recovery factor should also be a maximum. One disadvantage attends reservoirs approaching this ideal, namely that wells open near the oil/water contact will be subject to water-coning at high rates of production.

Since most sedimentary processes tend to segregate different sized particles into separate laminae, the isotropic reservoir is rare. The effects of grain orientation have already been mentioned. Other departures from isotropy may take the form of variations in permeability between layers (due for instance to grain-size lamination), or to the presence of partitions, generally sub-parallel with the depositional interface and consisting of thin layers of shale or other impermeable material, layers in which the sand itself is silty or argillaceous, or cemented, or carries abundant oriented flaky micaceous or carbonaceous (plant) material. Several authors have explored the effects of reservoir inhomogeneities on fluid production.[45,58,86,118] Polasek and Hutchinson[86] concluded that macroscopic fluid movement was controlled by the arrangement of sand and shaly sand, rather than by permeability variations within the clean sand of a reservoir.

The geometric arrangement of these inhomogeneities has a profound effect on the rate and completeness with which a reservoir can be drained. Two opposing geometries can be visualised, having either parallel or anastomosing permeability barriers, and each could be further subdivided according to their size and degree of continuity. In practice intermediate or mixed geometries prevail, but it can be seen that the more anastomosing and the more continuous the barriers, the greater the difficulty in draining the intervening sand, and the poorer the productivity and recovery factor.

One other very important departure from uniformity in sandstones needs to be mentioned, that is the fairly regular gradation in grain size (and therefore pore dimensions and permeability) from one boundary to another. When the gradation takes place more or less horizontally, it will obviously need to be taken into account in deciding well spacing. Vertical gradations in grain size are also common, and it is worth noting that where sands become finer upwards, hydrocarbons will tend to be concentrated in the worst part of the reservoir, and vice versa—a factor seldom taken into account in assessing the potential reservoir quality of sands deposited in different environments.

Sand bodies have been classified by Pettijohn et al.[82] in terms of their plan shapes. They are described as sheets, which are more or less equidimensional and more than about 5 km in extent; pods with length to breadth ratios of less than 3:1; and elongate bodies, having length to breadth ratios of more than 3:1. Elongate bodies are further subdivided into ribbons, dendroids (having tributaries) and belts (developed by fusion of several dendroids). It would, in theory, be possible to erect an elaborate three-dimensional classification, taking into account the shape in cross-section and the variations of internal organisation already described. Different values of reservoir quality might then be assigned to the different classes, and sands from natural environments fitted into the classification to give a rough quantitative basis for predicting the relation between reservoir properties and geology. However, such a complete classification would be cumbersome, many pigeon-holes would be under-employed, and few of the necessary parameters are as yet available. In particular, very few figures of porosity or permeability, and even fewer of water saturation, have been published in conjunction with identifiable depositional environments, and in those published the influence of cementation is generally unknown. For the present it may be more useful to consider combinations of shapes and other characteristics that are commonly encountered in oil fields, and to investigate their dependence on the depositional environment.

When the dimensions of sand bodies traceable in the subsurface in areas extensively explored for petroleum or coal are compared with one another, one is struck more by the similarities than the differences in the figures given for a wide range of shallow subaqueous depositional environments. Individual sand bodies identified by Conybeare[23] as channels (terrestrial and marine), bars, shoreline sands, and barriers have thicknesses ranging between 5 and 75 m, but the averages of each environmental group fall between 16 and 36 m; most lie between 20 and 30 m. There is a greater range in widths, from one to several hundred kilometres in extreme cases, the larger figures resulting from lateral migration and inconsistencies in definition, yet the average in each group is close to 4 km. Lengths are also variable, from under 5 km to 300 km, but again the averages for each group are similar to one another and seldom far from 50 km. Ratios of length to width are close to 10:1 for all these environments, except terrestrial river 'channels' where the ratio is about 50:1. Ratios of width to thickness tend to fall between 100 and 200 to 1. Although statistical analysis of a much greater quantity of data is desirable, it appears that prediction of the *detailed shapes* of fluvial and near-shore sand bodies is not likely to be the most impressive result to follow from the identification of their palaeo-environment. On the other hand, their directions of elongation and maximum permeability are important properties that can usually be predicted, as stressed by Selley.[96]

Examples of the Influence of Environment on Reservoir Character

With the preceding discussion in mind, some examples of the properties of different environments as they affect the nature and quality of sandstone reservoirs will be given in this section. A compilation of some significant features of sandstones formed in a variety of settings is given in Fig. 1.

Sand bodies of seven broad sedimentary environments or associations may be distinguished:[82] desert aeolian sands, alluvial sands, deltaic sands, coastal beach and barrier sands, inter-tidal and tidal sands, marine shelf sands and deep-sea sands. Deltaic and deep-sea sands are the subject of separate chapters in this book, and comment here will be concentrated on the other categories.

The attention that has been paid to the morphology, composition, and texture of the sediments of different environments has yet to be matched by investigations into their diagenesis. Yet there are hints that differences in the type and degree of cementation can be related to specific environments. The variations in porosity and permeability that attend these differences are at least as great as those for which variations in texture and detrital clay content are responsible, and the emergence of models predicting cementation is a necessary step towards the full application of sedimentology to petroleum exploration. Only one such comprehensive model is known to the writer; Fisher and Brown[32] relate carbonate and silica cementation to specific parts of the delta complex. Diagenesis is discussed in a later section of this chapter, mainly in relation to deep burial. But there is no hard and fast line between early and late diagenesis and, where possible, likely associations between cementation and the depositional environment will be noted in the present section.

Desert Aeolian Sands

These offer what are probably the most favourable of all clastic reservoirs—given adequate seals, trapping geometry, and petroleum sources, which are by no means inherent in this sedimentary association. Thicknesses can be enormous; as much as 300 m in dunes of the modern Sahara;[23] also an uninterrupted sequence of Permian Rotliegendes dune sands up to 200 m thick has been described in the Leman gas field of the southern North Sea.[117] These figures are matched by lateral extent; the dune facies occupies a belt 1000 km by 300 km in the Rotliegendes desert basin under the North Sea and The Netherlands. Such dimensions contribute, amongst other important factors, to the large reserves; in 1972 Glennie quoted the proved and probable reserves of gas from the fields of the southern North Sea, The Netherlands and Western Germany as nearly $2600 \times 10^9\,m^3$ (91 $\times 10^{12}\,ft^3$). These are believed to be predominantly aeolian facies, although with contributions from fluvial sands.[42]

Except at the unconformable base of a series, major accumulations of dune sand tend to be deposited on relatively flat surfaces, and likely modes of preservation also favour a flat contact with overlying strata. Since there is also a general absence of inter-laminated clay within aeolian deposits, the depositional geometry of dunes has little influence on the resultant sand reservoir. Traps are therefore usually structural. Where, exceptionally, the external form of individual dunes is preserved—as occurred in some cases during the rapid marine transgression of the Zechstein over the Rotliegendes in the North Sea[108]—the type of dune could control reservoir size and shape. Transverse dunes, for instance *barchans* (crescent shaped with the horns aligned down-wind), form at moderate wind strengths. Under stronger wind conditions *seif* dunes form and are elongated parallel with the prevailing wind; most southern North Sea basin examples are believed to be of the former type. A knowledge of the palaeo-wind direction is then important for predictive purposes, and may be available from regional studies or by examining oriented cores or dipmeter logs. In transverse dunes cross-bedding dips down-wind; in seif dunes most beds dip away from the axis at angles approaching $90°$.[42]

Grain-size lamination is common in aeolian sands and is a likely cause of permeability anisotropy.[117] Good sorting and rounding favour high primary porosity and permeability in dune foreset beds.

Cementation in the desert environment has been discussed by Glennie.[41,43] Being deposited above the water table, dune sands are less likely to be affected by early cementation than associated fluviatile facies. However, night dews alternating with high daily evaporation are thought to be responsible for 'case-hardening' of dunes, particularly by silica cement.[124] Secondary quartz cementation in Rotliegendes sands certainly varies more from grain to grain, for no apparent reason, than is common in sands of other environments, and this variability could be related to cementation above the permanent water table. Where the water table is close to the surface gypsum cement is likely, converting to anhydrite on deeper burial; later phases of anhydrite and carbonate cementation may occur if, as is often the case, the sands are associated with major evaporite deposits.[48] Despite the effects of later diagenesis, dune sands appear to provide better reservoirs than associated facies in the Rotliegendes of the

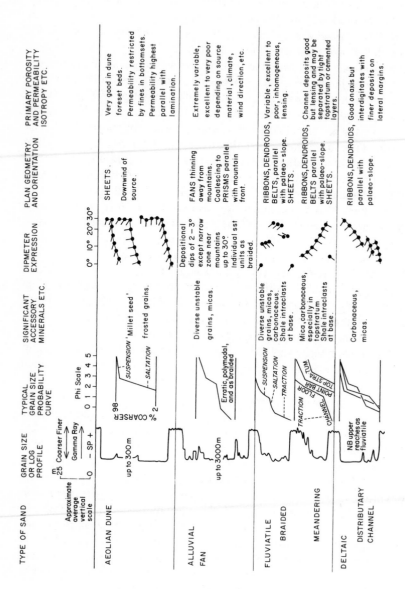

FIG. 1.

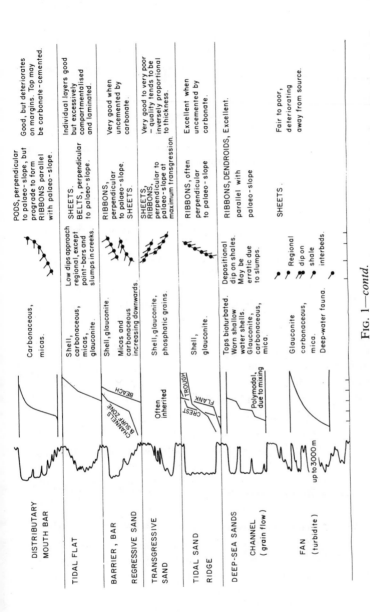

FIG. 1—contd.

FIG. 1.   Common reservoir sand types, with characteristic log profiles, grain size distributions, significant accessories, dipmeter expression, plan geometry and orientation relative to palaeo-slope, and reservoir characteristics. Compiled from Conybeare,[23] Glaister and Nelson,[40] Pirson,[84] Selley,[96] Shelton,[102] Visher,[117] van Veen[121] and others.

southern North Sea, according to Marie[73] and Blanche;[8] typical figures quoted are 25% porosity and 1000 md permeability.

Major deserts lie mainly in the trade wind belts between latitudes 10° and 30° north or south of the palaeo-equator, especially on the leeward side of continents and in the rain shadow of mountain chains. Important requirements for the accumulation of significant thicknesses of aeolian sand are first, an abundant source of material of the right size—usually provided by large river courses intermittently transporting debris from mountains sited to windward of the accumulation—and second, an appropriate rate of basin subsidence; this must not be so great as to invite premature termination by marine invasion.

Marine Shelf Sands

Although long believed to have great importance in the geological record, sands deposited on broad, open marine shelves have until recently been amongst the most poorly understood of sand facies. Modern counterparts are relatively inaccessible for detailed study, and today's shelves are in any case atypical because of the influences of Pleistocene glaciation. Evidence gleaned from modern and ancient examples has been summarised by Swift et al.,[112] Pettijohn et al.,[82] Selley,[96] and with particular relation to oil and gas fields by Conybeare.[23] A consensus on the processes involved in constructing large marine bars on open shelves has yet to be reached. Johnson[61] has recently developed a model based on three steps, the main constructional phase being attributed to storm conditions, with subsequent modification during fair weather varying in its effects from neap to spring tides. In contrast Hobday and Reading,[53] working on the same well-exposed Precambrian outcrops in Finmark, previously deduced that erosion occurred during storms, with deposition during quiet conditions.

On Recent shelves most attention has been paid to sand ridges—large, predominantly longitudinal, but occasionally transverse features up to 50 m thick, as much as 5 km wide, and 70 km long, deposited in depths of between 10 and 100 m under the influence of strong tidal currents. Elongation of longitudinal ridges tends to parallel the adjacent strand line. North Sea examples have flat bases and convex, asymmetric tops. Internal cross-stratification is little known, but includes large-scale inclined planar bedding. Internal discontinuities are present in some but absent from other ridges. There appears to be no progressive increase or decrease in grain size upwards. Preferential orientation of sand grain axes has been found to be at 45° to the strike of the ridges, parallel with the oblique movement of current ripples across their leading edges. The grains of sand ridges tend to be mature, well rounded, fine-grained and very well sorted. According to Conybeare there are no proven oilfield examples, but the sands should make excellent reservoirs. Calcite cementation seems a possibility, however, where lime grains or calcareous organisms are abundant.

Although Conybeare deduces that grain size neither fines nor coarsens upwards, the ancient sub-tidal bars studied by Johnson and by Hobday and Reading cited previously, and Recent ridges of the New Jersey Shelf studied by Stubblefield et al.[111] all coarsen upwards. Sedimentary structures and depositional current directions also change from base to top.

Presumably many mature quartzose sheet sands of the past (such as the Ordovician St Peter Sand of North America) were deposited on stable cratonic shelves by repeated migration of sub-tidal bars and also various coastal sand bodies, described in the next sub-section, the elements of which may often be too inextricably mixed for individual identification.

Sheets and lenses that formed during transgressive and regressive phases are characterised by upward fining and upward coarsening grain sizes, respectively.[23,119] Transgressive sands tend to be widespread but thin. Consideration of the depositional process suggests that the slower the transgression and the more thoroughly the sand is winnowed the more likely it is to be carried shorewards to accumulate in quantity only at the ultimate shoreline, whereas rapid transgression may leave behind a thicker deposit which, however, is likely to be less well sorted. According to Oomkens[80] the transgressive sands in the Rhone delta complex, for instance, seldom exceed 2 m thick. This simple picture does not take into account the point that many marine transgressions occur across coastal plains, and sands previously deposited under alluvial, dune, or beach conditions become incorporated into the transgressive sheet with more or less reworking depending on the rapidity of the transgression, the bottom slope and wave fetch. Thus, ancient transgressive sands listed by Conybeare[23] are by no means all thin, and some have excellent reservoir properties. They vary from 6 to 90 m thick, have porosities of 15 % to 28 %, and permeabilities of 80 to 3000 md. Glauconite, phosphatic nodules and relict faunas are common in transgressive sands. Because they fine upwards and grade into marine shales above, their reservoir properties tend to be poorest at the top; perhaps because of this, and despite the favourable relationship for trapping pointed out by MacKenzie,[72] fewer commercial oil and gas fields are recorded in transgressive than regressive sandstones.

Broad sheets of regressive sand, on the other hand, can only develop where there is a continuous supply of sand, for example in the vicinity of deltas. Therefore, although the sand is deposited in the high-energy environments of beaches and barrier bars, it may incorporate relatively large proportions of river-derived immature sand grains (particles of shale, schist, feldspar, chert, micas, limestone, and carbonaceous matter) which can break down mechanically or chemically after burial to reduce the initially good porosity and permeability. However, rapid deposition favours freedom from early calcite deposition. Reservoir properties tend to be best near the top of the sheet which was deposited in the highest energy conditions, possibly incorporating beach and dune facies.

Beaches, Barriers and Bars

When preserved as distinct entities, probably as a result of rapid immersion, sand bodies directly related to and paralleling the palaeo-shoreline form an important group of potential reservoir sands. Emphasis is placed on different elements by various authors. Pettijohn *et al.*[82] stress the beach as a key formative feature of barrier islands. Selley[96] emphasises the difficulty—and perhaps the irrelevance—of distinguishing barrier islands from offshore bars, spits, and tombolas in subsurface studies. The features which are important in this regard are their linear form; a characteristic thickness (determined by tidal range and depth of water above the wave base) commonly between 10 and 20 m, a width of the order of 4 km, and dependence on wind-driven wave energy as the chief source of erosion, transportation and deposition of sand. Wave dominance leads to a general upward and shoreward coarsening of grain size, with the most mature, cleanest and well-sorted sand at the top. Aeolian dune sand may cap the units.

Beach deposits are characterised by low-angle lamination or thin bedding dipping gently seaward; the bedding planes are often marked by concentrations of heavy minerals. Bed thickness tends to increase upwards. Seawards, the sands fine into bioturbated and lenticular (flaser) bedded silty and dirty sands of progressively diminishing reservoir quality, and finally into marine shales. Landwards, beaches may be bounded by coastal marsh and alluvial or other terrestrial facies. If a lagoon is present it may be represented by muddy or sandy deposits depending on the tidal range. Where large volumes of water are exchanged between lagoons or estuaries and the open sea, tidal channels become important and are responsible for steeper forms of cross-bedding, dipping in a variety of directions. An upward-coarsening, vertical succession of textures and associated sedimentary structures, created by lateral migration of tidal inlets, has been described by Kumar and Sanders,[67] and is believed to contribute significantly to the construction of ancient barriers. Whether a complex has grown mainly through shore-face processes or inlet migration, the high energy of the environment in the shallow, inter-tidal and supra-tidal zones seems likely to ensure an absence of significant subdivision by clay or shale partings. Sand grain long axes have been found to trend at right angles to the shore in Recent examples,[25] and permeabilities should be highest in this direction also. This would appear to be statistically likely for both modes of accretion mentioned above.

No generalisations about early cementation seem possible as yet, except that carbonate cement is more likely to be formed the warmer and drier the palaeo-climate and the slower the rate of sediment movement. Even in temperate climates beach ridges often consist of aeolian lag concentrates of shells which, particularly if they are aragonitic, may provide a local source of carbonate cement for adjacent beds. Many shoreline sands, nevertheless, remain poorly cemented and provide important oil and gas reservoirs, which have been particularly well-documented in the literature relating to North America. Conybeare[23] reviews eighteen examples of reservoirs in barrier and similar bars; they have an average thickness of 20 m, width of 4·4 km and length of 42 km. Porosities are mainly in the range 15–25%, and permeabilities mostly between 50 and 2000 md.

The Piper oil field, with recoverable reserves of 650–900 million barrels, is of interest as an example of a northern North Sea Jurassic shoreline reservoir. It is described by Williams et al.[126] as a series of stacked and possibly imbricated barrier-bar and other littoral shallow shelf marine sand bodies, separated only by thin and largely discontinuous siltstones and silty shale. The gross sandstone thickness varies from 35 to 110 m and is made up of transgressive and regressive units ranging from 2 m to 30 m thick. Porosities are between 22 and 27% (average 23%) and permeabilities between 150 md and 10 d (arithmetic average 1495 md, geometric average 761 md).

Tidal Flat and Estuarine Sands

Estuaries and tidal flats are dominated by reversing currents. Although muddy sediments tend to predominate, especially on tidal flats, maximum current velocities aided by wave movement are usually competent to transport sand.[82] Finer material is deposited from suspension as current velocities pass through zero with each tide reversal, and is not easily picked up again as velocities increase. Thus, there is a built-in tendency towards the accumulation of sand/mud laminae. Because estuaries and tidal flats are protected from the open sea, wave and current energy tend to be damped out as sediments build nearer to high water level, and both environments are therefore represented by upward-fining sequences.

In estuaries, mutually evasive ebb and flood channels form a variety of sizes of shifting sand waves elongated parallel with the current direction, characterised internally by cross-bedding (often bi-polar), clay drapes, and ripple bedding. Thicknesses are limited by water depth. Tidal sands in some estuaries are moderately clean and well-sorted, and could provide fair hydrocarbon reservoirs. Others may be too argillaceous or silty, subdivided by clay laminae, or suffer too much from later compaction and diagenesis as a result of an abundance of immature sand grains.

Tidal flats frequently border estuaries and the lagoons behind barrier islands, and form extensive parts of the sub-aerial margins of deltas. A selection of modern and fossil examples has been summarised by leading authorities under the editorship of Ginsburg.[39] Despite differences in detail it is usual to find an upward increase in the proportion of mud inter-laminated with sand in the inter-tidal zone, culminating in muds trapped in vegetation on the salt marsh. Herring-bone cross-stratification, wavy and lenticular bedding, flaser bedding, small-scale ripples and bioturbation are common and combine to give a sand body excessively partitioned by parallel and anastomosing permeability barriers, especially in the top few metres. Furthermore, the tidal flat is frequently cut by creeks which deposit upward-fining sequences of sand and inter-laminated sand and mud on the inner banks of meander loops. Slumping is also common. The complex inhomogeneity resulting from these depositional processes would, in a hydrocarbon reservoir, appear to be inimical to efficient oil or gas drainage, even though individual sand layers might be well-sorted, clean and uncemented.

Fluviatile Sands

Because of their accessibility and importance to civilisation modern rivers and their flood-plains have probably received more attention than any other sedimentary environment; yet despite (because of?) this, there is often misunderstanding about the nature of fluviatile deposits preserved in the geological record. It is necessary to distinguish clearly between three things: the geometry and distribution of the facies in and around a river over a short period of time (a few years); the sedimentary sequence produced as the stream moves laterally over a longer period (a few to several hundred years); and the results of the continuation of the same processes over thousands or millions of years during which fast or slow, steady or intermittent subsidence may take place. Sedimentologists studying the second of these aspects have deduced the familiar upward-fining alluvial sequence. The third results in the building-up of a thicker series, the character of which varies according to rates of subsidence, changes in sediment supply, and stream migration.

Sands may be deposited from two distinct fluvial regimes—braided and meandering streams. Braided streams tend to characterise conditions of higher gradient, stronger but more intermittent discharge, and coarser sediment size than meandering streams. They generally occur nearer a river's source and frequently in less temperate climates; transitions from one regime to the other, however, are not uncommon. Until recently more studies relevant to geological problems had been published about meandering streams than braided ones. Models of the former are given by Allen,[1,3] and Visher,[120] and quoted in many texts.[82,87,96] According to these authors and most other workers in this field, the fining-upward sands of meandering streams are formed mainly by lateral accretion on point bars on the inner side of meander loops, whilst erosion occurs on the outer side.

In braided streams, on the other hand, longitudinal braid-bars elongate downstream during flood and grow by lateral accretion during low river stages; old channels periodically become choked and divert flow to new courses. Earlier descriptions of braided streams[28,30,92] concentrated on examples with pebbly deposits, but more recently, Cant and Walker[16] have proposed a model for sandy braided systems. Although the authors are at pains to demonstrate differences from the deposits of meandering streams, one is led to the conclusion that despite the operation of different mechanisms the two types may be difficult to distinguish in boreholes. Both systems produce a fining-upward sequence with a thickness comparable to the river depth during flood. Both commence, above a scoured base, with coarse sandy sediment in which mudstone intraclasts may be abundant, and pass up into planar- and cross-bedded channel sands, later into finer sand, silt, or mud, characterised by smaller-scale planar- and cross-lamination and ripples. In complete sequences both are succeeded by silts or clays deposited from suspension; somewhat similar fine-grained deposits also occupy abandoned channels. The differences between sediments of the two regimes lie mainly in the importance of fine-grained flood-plain and channel-fill deposits (all of which are more abundant in meandering river systems), and in the presence in the middle of the braided sequence of large-scale planar-tabular cross-beds of somewhat coarser sand showing flow directions at large angles to the trough cross-bedding immediately below and above. These distinctions are difficult to make in cores.

In ancient rocks the thickness of flood-plain deposits that is preserved may be determined less by the river regime than by rates of subsidence. When subsidence is sufficiently rapid the whole of the deposited sequence may be preserved, but when subsidence is slow, stream migration will erase the upper parts of underlying sequences, and the slower the subsidence the less likely it is that any flood-plain deposits will survive. Not only does slow subsidence make it more difficult to distinguish braided from meandering stream deposits, but it may also lead to accumulations consisting predominantly of the lower parts of upward-fining sequences—in extreme cases merely beds of conglomerate. It may be noted in passing that subsidence is rarely likely to be so regular that erosion proceeds uniformly to a particular level in the depositional sequence. In general some areas will subside more and preserve higher parts of the sequence than elsewhere. Since the finer deposits of the upper parts of the sequence compact more readily, these areas will tend to be self-perpetuating—at least until they attract and temporarily trap a migrating channel.

The processes described above lead to the building of tabular masses of sand extending widely across the alluvial plain, with thicknesses limited only by sediment supply, subsidence, and time. Channel trends, cross-bedding, grain orientation, and maximum permeability will all have their preferred direction parallel with the palaeo-slope. Channel and point-bar sands may be clean and well-sorted, with good primary porosity and permeability, but these sands are likely to be lenticular.[96] Reservoirs formed from them will tend to be subdivided by poorly-permeable layers represented by the tops of individual upward-fining sequences when these are preserved, and in some arid climate cases by carbonate-cemented beds formed during early diagenesis in the basal part of each sequence.[113] Lateral and vertical inhomogeneity is likely to be greater, and the grains less mature and less well-sorted in braided than in meandering stream deposits.

It is noteworthy that a somewhat different picture emerges from the literature on fluviatile sands as hydrocarbon reservoirs (see for example reference 23), which tends to concentrate on sinuous sand-filled channels embedded in less-permeable facies. In many cases channels were identified by their shape in plan, often after much drilling. They frequently appear to represent broad meander belts rather than individual river courses. In nearly half the cases described by Conybeare[23] these 'channels' are cut down (incised) into older rocks. This must represent a comparatively rare form of preservation, and it is suspected that many fluviatile sheet-sands have gone unrecognised because they lacked obvious channel morphology. Conversely, the sand filling many sinuous channels may well have been deposited by some other means, for abandoned meander loops in active flood-plains are usually filled with mud, except for a plug of sand at each end.[1] The channel sands quoted by Conybeare range from 6 to 75 m thick (average 36 m), 250 m to about 20 km wide (average 4 km), and were traced for distances of 6 to 160 km. Porosities ranged from 10 to 25 % (average 17 %) and permeabilities from 60 to 2000 md (arithmetic average 500 md). According to Morgridge and Smith,[78] most of the recoverable hydrocarbons of the Prudhoe Bay field, Alaska, occur in braided stream deposits of the Permo-Triassic Sadlerochit Formation. Porosity, restricted by secondary quartz cement, ranges from 20 to 24 %, and permeability from 300 md to several darcys.

The sands of delta distributary channels, described in detail in Chapter 6, have quite different characteristics. They tend to be relatively straight and build vertically rather than migrate laterally, in response to subsidence of the delta platform. Thicknesses up to 75 m are not uncommon. Prograding seaward, distributary channel sands tend to display an upward-coarsening profile. They provide the prolific 'shoestring sand' hydrocarbon reservoirs of many parts of the world.[23]

Alluvial Fans

Coalescence of alluvial cones or fans gives rise to a linear wedge of coarse detritus running along the foot of a mountain chain. When the junction between mountains and plain is an active fault zone, as is often the case, the wedge can reach a thickness of 3000 m or more. an important exception to the sheet-like geometry of most clastics. Scree slopes dip at 30° or so, but depositional dips rapidly diminish down-slope to 2–3° as fluviatile (mainly braided) transport takes over from gravity sliding on the major part of the fan. In tectonically more stable regions Selley[96] and others have advanced the view that migrating fans can also produce extensive sheets of detritus by the erosion of retreating scarp lines.

Deposits are extremely variable, varying from boulders to clay, often intermixed as a result of mud flows. Grain size diminishes and sorting improves in the direction of transport. Composition is extremely immature and reflects the lithologies of the eroded uplands with little modification. The primary porosity and permeability, which may be extremely high in sand and gravel lenses free from fines, is therefore particularly vulnerable to compaction and diagenesis on burial. Relatively early ferruginous and carbonate cementation also seems to be common. In an interesting account of the cementation of alluvial fans in southern Nevada, Lattman[68] states that it is at its greatest on fans composed of carbonate and basic igneous debris, less on fans built of siliceous sedimentary detritus, and least in those constructed of fragments of acid igneous rocks. The best-developed cementation is in poorly sorted layers of alluvium containing more than 25% of material coarser than pebble size. A major source of calcium carbonate is apparently wind-blown silt and sand. Consequently, fans of non-carbonate detritus are best cemented down-wind of playas high in precipitated carbonates.

Cretaceous alluvial fans buried by marine Cenomanian shales provide important oil reservoirs in the Sirte Basin of Libya at depths of between 2500 and 4500 m. They include the giant fields of Zelten, Sarir, and Waha.[23] In view of Lattman's findings it is interesting to note that these largely uncemented fan deposits overlie (and were presumably derived from) predominantly granitic basement.

SANDSTONE DIAGENESIS

Mineralogical changes that take place in sands after deposition (diagenesis) are of importance to the petroleum geologist in so far as they modify the porosity, permeability, entry pressure and irreducible water saturation. Since buried sandstones are under a confining pressure, due to the overburden, which usually exceeds the fluid pressure in the pores, there is a strong tendency, according to Le Chatelier's Law, for any changes that occur to be accompanied by a reduction, rather than an increase, in porosity. This situation may be locally reversed beneath unconformities, where exhumation and weathering may have led to dissolution (mainly of carbonate cements) and enhancement of porosity and permeability. Furthermore, it has lately been proposed by Schmidt and McDonald[93] that early carbonate cements can be dissolved during subsequent burial, as a result of the release of carbon dioxide during decarboxylation of organic matter in adjacent source rocks.

Apart from simple crushing of weak grains, reduction in pore volume may result from the *addition* of minerals from outside (cementation proper), or through the *loss* of material forming the sand grains by dissolution at their contacts, leading to compaction (i.e. pressure solution); in some circumstances compaction may be accomplished by material dissolving at grain contacts and reprecipitating in the pores (solution transfer). Cementation proper is believed to be the most important of these processes. Except in rare instances where cementation is restricted to the vicinity of grain contacts, all of these volume changes lead automatically to reductions in pore dimensions and permeability, and an increase in entry pressure; inasmuch as the internal surface area is increased, a slight increase in irreducible water content is also indicated. Certain mineralogical changes, however, can also affect entry pressure, permeability, and irreducible water saturation without necessarily being accompanied by appreciable change in porosity. This effect arises from the modifications of pore size and shape related to the crystal habits and the surface chemistry of different minerals. The principal examples are alterations involving clay minerals; most, but not all such changes lead to poorer reservoir properties.

Almost any mineral stable in the temperature range 0–250 °C and at pressures up to about 25 000 psi ($172 \times 10^6 \, Nm^{-2}$), can develop in reservoir rocks, given a source of the necessary ions (either from unstable minerals within the sand, or pore fluids initially present or passing through it), favourable reaction kinetics, and the time to develop. Many have been recorded, but only a limited number occur commonly in sufficient quantity to be important to the petroleum geologist as cements. Of these, quartz, and the carbonates calcite, dolomite, and to a lesser degree siderite, cause by far the most serious loss of porosity; in special circumstances anhydrite, barite, fluorite, halite, iron oxides or pyrite may also be important in this regard. Secondary feldspar often accompanies secondary quartz and can be volumetrically important in some sandstones. Of all the above, calcite, dolomite, and secondary quartz are by far the most common. The clay minerals kaolinite (or its close relative dickite), illite, chlorite, montmorillonite, and mixed-layer species such as corrensite have been demonstrated to have grown authigenically in sandstones and often have significant effects, particularly on permeability, entry pressure, and irreducible water. Of these minerals, kaolinite, illite and chlorite are commonest.

In general one would expect that those particle shapes or sizes which resulted in greater surface area for a given volume, would lead to lower permeability and higher irreducible water saturation for a given porosity, and would also probably be responsible for a lower hydrocarbon recovery factor. Authigenic clays occur either as aggregates of crystals packing pores, or as oriented coatings around sand grains. Particle size is usually under 5 μm, but is occasionally as much as 15–20 μm. Pore-filling authigenic clay may be widely dispersed, indicating crystallisation from ions carried for some distance in the pore water, or concentrated in place of particular precursor grains, as is the case with kaolinite produced by *in situ* alteration of feldspar, which provided the necessary alumina and silica. Grain coatings may have crystals arranged with their long dimensions either parallel with, or perpendicular to, the surface of the host grain. Clay coats can be particularly important as inhibitors of later quartz cementation.[52] Kaolin forms pseudo-hexagonal plates, often regularly arranged in book-like or vermicular stacks.[101] Chlorite shows slightly more angular radially-oriented plates, and can also form dense aggregates of needles in pores.[17] Illite occurs either as aggregated interlocking plates or as fine fibres, and similar habits have been reported for montmorillonite, although it is usually finer.[69] Corrensite forms networks of crinkled or folded sheets ('cornflake' structures[4]).

A practical illustration of permeability/porosity relationships to clay mineralogy is given by Stalder,[110] who showed that in Rotliegendes dune sandstones with authigenic kaolinite and illite, those sandstones having characteristic kaolinite booklets gave higher permeabilities than those containing the finer and more thread-like crystals of illite. Effects on permeability were significant, even though total amounts of clay were only about 5%.

It is often difficult to distinguish between authigenic and detrital clay in sandstones, particularly under the light microscope. Authigenic clays are generally more homogeneous, better crystallised, coarser, and may show one of the forms of regular organisation described above. The latter is not an infallible guide on its own, however; Walker and Crone[122] have described tangential and radial coatings of clay on grains which have resulted, along with random aggregates in pores, from mechanical infiltration of clay into coarse bands of desert alluvium shortly after deposition. These infiltrated clays usually exhibit some geopetal relationships, however, which are not shown by authigenic clays.

A profusion of papers on the fundamental aspects of cementation and other forms of sandstone diagenesis were published in the late 1950s and during the 1960s. In general, much careful and conscientious work of this period was carried out too early to receive the full benefit of the advances that were concurrently being made in the identification of sedimentary environments. Consequently, it is often difficult to isolate and correctly attribute the diagenetic effects of the environment of deposition from those due to later conditions—an essential preliminary to the deduction of general principles. An excellent review of the subject up to 1972, summing up petrographic, physical and chemical aspects, is given by Pettijohn et al.[82] Since then results stemming from the application of scanning electron microscopy have tended to dominate the literature, concentrating attention on the clays, which had previously been difficult to study, but including interesting investigations into the development of quartz overgrowths.[85,123] In the last few years data gathered during exploration in the North Sea, The Netherlands and Germany have begun to be released and are contributing further ideas on the relation of diagenesis to burial, temperature gradients, and hydrocarbon accumulation.[100]

CEMENTATION—THE PROSPECTS FOR PREDICTION

The ability to understand the observed variations in sandstone cementation in terms of the history of the rock after burial is obviously of great academic interest, but to the petroleum geologist the over-riding objective must be, so far as possible, to *predict* the types and quantities of cements that are likely to be present in a given sandstone in order to allow for their effects on reservoir properties. Prediction is required on two levels. First, before drilling, the geologist needs to be able to indicate to management the probability that a sand in a given regional setting and at a given depth will have attractive reservoir properties. Second, when a reservoir sand has been located and identified by drilling, he would like to be able to specify to closer tolerances what changes are likely to occur in untested parts of the reservoir, or in adjacent sands. These are daunting tasks; there is a tendency in some quarters to regard the effects of diagenesis as so capricious that it is fruitless to attempt to quantify them, other than as an element of random variability. Yet modification of the depositional pore space by diagenesis is often so great that unless the attempt is made there is little advantage in applying sedimentology to sandstone exploration.

A statistical approach seems appropriate to the problem of predicting porosity in advance of drilling, particularly in the case of deep prospects. There is good evidence from various parts of the world that sandstone porosity tends to diminish with increasing depth. In rocks of the same geological era from a given area, the relationship is remarkably linear. Atwater and Miller[6] studied 17 367 analyses of conventional cores from 101 fields and many wildcats in southern Louisiana, mainly from Miocene and younger sandstones, and found an average reduction of 1·265% porosity per 1000 ft of burial (4·15% per km). Maxwell[76] compiled data from a variety of sources, involving over 8000 observations from formations ranging from Ordovician to Miocene (though mainly Tertiary), and found that gradients of porosity reduction were greatest when temperature gradients were highest. Gradients varied from about 1% per 1000 ft (3·28% per km) for Oligocene sands at a temperature gradient of

7 °C per 1000 ft (23 °C per km) to 2·56 % per 1000 ft (8·4 % per km) for sands of similar age at a gradient of 10 °C per 1000 ft (32·8 °C per km). Porosities also tended to decrease with age.

Using data from the North Sea, Selley[98] has quoted gradients of 3 % per 1000 ft (9·84 % per km) for Tertiary sands, 2 % per 1000 ft (6·56 % per km) for Jurassic sandstones and 1·5 % per 1000 ft (4·92 % per km) for Permo–Triassic sandstones.

It should be noted that the depths on which the gradients quoted are based were in all cases *present* depths. In the case of the Gulf Coast, with its history of continuous subsidence since the early Mesozoic, these may approximate *maximum* depths. In the North Sea, with a far more complex history, this is by no means the case.

Gradients can either be based on the *mean* porosity of a number of samples at each of a series of depths, or on the *maximum* porosity observed at those depths. Use of the *mean* (employed by Atwater and Miller, and also Selley) has the statistical attraction that it takes all the observations into account but, equally, suffers from the objection that it may be influenced by samples in which primary matrix limited the porosity before burial. The *maximum* porosity (preferred by Maxwell) is statistically imprecise, especially when the quantity of data is small, but is more easily interpreted geologically. The range of porosities at any depth is considerable, and the comparatively uniform gradients observed are the result of the interaction of a large number of factors. They are not, as is sometimes mistakenly assumed, entirely—nor even principally—the result of pressure solution promoted by overburden pressure, but involve all forms of cementation. They are also influenced by the primary depositional porosity, the size distribution, shape, mechanical properties and chemical stability of the original constituents, and by the nature and pressure of the pore fluids. This complexity was visualised by Maxwell[76] and many aspects of it have been confirmed theoretically and experimentally by other workers since.[9,10,29,88,109] Many of the factors are influenced by the depositional environment. Selley[98] has taken this into account by constructing families of porosity/depth lines representing different environments, and has gone a step further by associating expected permeabilities with each. The result is a system of prediction which needs only one refinement—the addition of confidence limits—to approach the ideal of enabling a probability figure to be attached to the reservoir properties of an undrilled target located by seismic surveying.

The problem of predicting changes in permeability with depth has also been tackled by Hsu.[59] Working on deep-sea sandstones buried to about

4 km in the Ventura field of California, he has deduced the empirical relation:

$$k = Cd_m^2 \exp(-1 \cdot 31\, \sigma_\phi)$$

where k is the permeability in millidarcies. C is an empirical number, d_m is the median grain size in millimetres, and σ_ϕ the sorting expressed as phi standard deviation. Another factor, the 'granular parameter', G, is defined as:

$$G = d_m^2 \exp(-1 \cdot 31\, \sigma_\phi)$$

The granular parameter is a measure of the suitability of a sand as a hydrocarbon reservoir *before* any diagenetic changes. The empirical number C gives a measure of the degree of compaction; an increase in permeability is directly proportional to C. By plotting measured permeability against granular parameter for groups of samples from the same productive zones, a family of roughly parallel straight lines is obtained, one for each depth zone. For any given region, a general knowledge of depositional environments should indicate a range of granular parameters. After the effect of compaction has been defined empirically by the above method, the permeabilities of types of reservoir sands not yet encountered in explored zones can be estimated.

A more deterministic approach based on a better understanding of the detailed physico-chemistry, thermodynamics, and petrology of mineralogical changes seems likely to find its main application in those cases where the transfer of ions between the sandstone and outside sources is a minimum, such as the alteration of feldspars, mafic minerals, and volcanic glass to clay minerals, and of one clay mineral to another. In such cases it may be possible to specify the initial solid phases (mineralogy) and the main chemical features of the initial pore fluids (determined by the depositional environment), and to predict the rates of change and the products under specified conditions of temperature and pressure (burial). The main features of several common reactions are now generally agreed. Thus, kaolinisation of feldspar[82] requires pore waters low in dissolved solids, including K^+ and H_4SiO_4, and having relatively low pH, probably determined by dissolved H_2CO_3. Such conditions are most readily met in shallow meteoric groundwaters, and most feldspar kaolinisation is thought to occur either soon after burial, or during later uplift into the weathering zone, or adjacent to faults or other channels which provided access for suitable waters. Kaolinite changes to other clays such as illite at temperatures above 150 °C in the presence of alkali ions. Montmorillonite dehydrates between temperatures of 110–150 °C and takes up magnesium or potassium to form mixed-layer clays, chlorite or illite. In the Gulf Coast, USA, this change has been demonstrated in shales between depths of about 5000–12 000 ft.[57] Illites and chlorites are relatively stable, except under acid conditions, and increasing temperature brings about only a higher degree of crystallinity; they can therefore be used as geothermometers. Almon *et al.*[4] have found that diagenesis of a volcaniclastic Cretaceous sandstone in Montana was controlled by the depositional environment, with corrensite (accompanied by calcite and dolomite) restricted to samples from delta distributary channels and mouth bars, and montmorillonite plus calcite characterising bay-beach, crevasse-splay, lagoon, barrier, island and shallow sub-tidal environments. The authors were able to deduce the chemistry of the pore waters involved in the transformations fairly closely. Thomas[114] has similarly found a close correspondence between mineral paragenesis and thermodynamic prediction in certain fine-grained brackish to fresh water Rocky Mountain Cretaceous sandstones.

Difficulties arise immediately cementation involves significant exchange of material with formations outside the reservoir sandstone under consideration. Unfortunately such cases are probably the most important. Solubilities of common cementing minerals under realistic subsurface conditions are only of the order of a few parts per million to parts per thousand; consequently, large volumes of water must pass through a sand to effect significant changes in porosity. Equally, much water is expressed from associated fine-grained rocks and from more deeply buried sandstones during compaction or thermal dehydration, and during its escape towards the surface is attracted to the most permeable pathways. It follows that in most sands, even where lenses are isolated in thick shales, the average pore water chemistry over a period of time will tend to be dominated by ions acquired *outside* the sand.

Given an adequate knowledge of the geochemistry, the difficulty in predicting the amount of cement likely to be deposited in a given sand results from the impossibility of knowing the flow of water through it from various sources during the course of its geological history. Qualitative predictions can, however, be made in reasonably simple cases. Thus, a sand situated down the pressure and temperature gradient from a thick marine shale sequence buried sufficiently for the montmorillonite–illite transition to take place might be expected to show secondary quartz cementation, since the clay reaction involves release of silica,[57] and the solubility of quartz is strongly related to temperature. Moreover, since almost every quartz grain in a sand is a potential nucleus, and since solutions supersaturated with respect to quartz are stable over long periods (and crystal growth of the mineral is very slow), it might also be expected that roughly equal thicknesses of quartz would be deposited around most sand grains. This is, indeed, in accordance with petrologic observation of marine sandstones of different ages from various parts of the world. If the shale originally contained much calcareous matter or if the sequence included significant amounts of limestone, calcite cementation could be expected. Shinn *et al.*[104] have recently shown that, contrary to general opinion, lime muds probably compact appreciably during early burial, whilst Hancock and Scholle[47] have demonstrated that the chalk under the North Sea compacts steadily by pressure solution with increasing depth. Analyses by Hower *et al.*[57] indicate that calcite is lost progressively from Gulf Coast shale down to about 4 km. These examples suggest that substantial amounts of water laden with calcium and carbonate ions are released from limestones and shales at various periods in their genesis. Unlike quartz, however, the solubility of calcite decreases with temperature whilst increasing with pressure; Arntson[5] and Sharp and Kennedy[99] deduced from experimental studies that solutions saturated at depth will become less saturated as they travel along normal geothermal and hydrostatic pressure gradients towards the surface, and hence that subsurface precipitation is likely to be confined to situations where a substantial pressure drop occurs, as in fractures and joints. One might also infer that sands would be most likely to become cemented by calcite where they were in contact with compacting (preferably over-pressured) shales and limestones, and where they provided a low impedance path to zones of lower pressure. This may help to explain why calcite cement in sands is frequently confined to the vicinity of shale contacts.

A hypothesis commonly offered for the latter phenomenon involves ionic filtration,[35] treating shale as a semi-permeable membrane and requiring fluid pressure in the sand to be higher than that in the shale. Whilst there may be cases where this mechanism applies, it is a poor explanation of those common examples in which sands on *both sides* of a thin shale are cemented in contact with it. There are, in fact, likely to be many different reasons for the local cementation of sands by calcite, which is much more variable in its distribution than quartz cement. Not least of these is the question of nucleation. Unlike quartz, again, every grain in a sand is *not* a nucleus for calcite growth. Supersaturation with respect to calcite has therefore to be high before precipitation occurs, unless calcite nuclei (such as scattered skeletal fragments, derived calcareous grains, earlier precipitates from local reactions, etc.) are present. The depositional environment may therefore assume special importance in determining the distribution of calcite cement, for not only may redistribution of abundant shell debris and other primary calcium carbonate lead to local cementation during early diagenesis in some marine environments, but quite small amounts of entombed carbonate could act as seeds later to encourage differential cementation from migrating waters.

It can easily be seen that although qualitative predictions of cementation can be attempted from a knowledge of the stratigraphic succession, together with broad generalisations about the pattern of fluid flow, attempts to quantify these ideas are bound to be frustrated by the inability to define the detailed plumbing of the system as it varied from the time of deposition onwards to the present, for example as a result of differential tilting movements and faulting. Although the principal faults can be detected by seismic surveys, many smaller ones may be undetected, and movements have not always been solely in the direction indicated by the present geometry.

Another empirical approach to the problem of porosity prediction has been employed by the author.[113] Based on detailed petrographic measurements, it is possible when samples from the reservoir are available. It exploits the serial nature of diagenesis and the smoothing effect of successive phases to give an indication of the *maximum* porosities to be expected.

If a suite of permeability/porosity plugs from a typical sandstone reservoir is analysed quantitatively in thin section by point-counting, correlation of individual components with porosity or permeability is often disappointingly low and rarely statistically significant, except in cases that are so simple that the correlation is trivial. This tends to be so even when there is good petrological evidence that the distribution of some components is controlled by others. In a typical sandstone the important parameters might include average grain size, shape, and sorting, packing density, proportions of three or four grain types of differing stability, amount of primary matrix, one or more carbonate cements, authigenic clay, secondary quartz, and visible porosity—at least a dozen factors. Quite apart from the difficulty of determining some of these objectively (plus the sampling error attached to individual percentages for so many components, inevitable unless rather large numbers of points are determined), the number of interacting factors is too great to give clear-cut answers, even with the most sophisticated statistical analysis.

Examination of many sandstones from various geological systems in different parts of the world, however, suggests a common pattern which has quantitative implications. Three stages of porosity reduction are commonly discernible, exhibiting the following characteristics:

(1) Early cements, developed at no great depth, and fairly directly related to the depositional environment, tend to be variable in their distribution.

(2) Later cements, developed during deeper burial (except those introduced along and concentrated close to faults) are more uniform in their distribution—which is, however, necessarily limited by that of matrix and early cements.

(3) When developed, compaction (mainly by pressure solution) tends to be later than the main phases of cementation, and is most pronounced in those parts of the rock where the grains are not supported by cements or matrix.[51,105]

There is frequently some overlap between these different stages and there may be still later phases of cementation, but this need not alter the overall picture. It will be noted that each stage places restraints on the degree to which subsequent processes can operate, and that as a result the later, more universal processes tend to iron out earlier differences in porosity. Take, as a simple example, a regressive bar sand, clean towards the top but containing variable amounts of detrital clay matrix lower down. During early diagenesis, calcite cement might be concentrated near the crest by calcitisation of aragonite shell debris. During burial to between 2 and 4 km, quartz cement deposited from migrating fluids may tend to form overgrowths of relatively uniform thickness around quartz grains not protected by clay or calcite. On deeper burial, pressure solution is concentrated where there is least mechanical support—very often in those places where secondary quartz was inhibited by minor amounts of clay.

The principle can be restated more generally in terms of a diagram (Fig. 2) in which packing density is plotted against the sum of the pore-filling components (clay, carbonates, secondary quartz, etc., but excluding visible porosity). If there is more than about 40 % matrix and cement (point A) the grains float; the line to the right of this point descends at 45° to zero packing density representing pure shale or limestone (B). If there is less than 40 % matrix and cement, a sandstone showing no compaction would be represented by points scattered around a straight line parallel with the abscissa (AC). If, on the contrary, compaction is only limited by the solid components occupying the pores, points would plot on a continuation of the 45° zero porosity line BA to 100 % packing density (D). Typical sands would be expected to show intermediate situations, with the slope of the

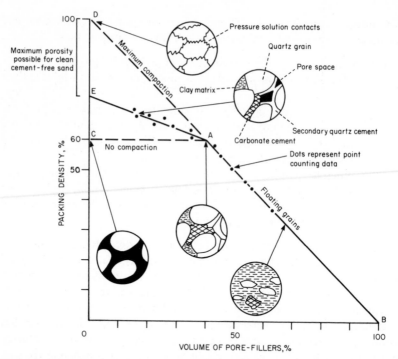

FIG. 2. Relationship between packing density and volume of pore-filling matrix and cement for typical sandstones. The slope of the line AE depends on the subsurface conditions to which the sandstone has been subjected. Insets show possible textures and compositions.

line AE determined by the extent of compaction, as shown on the hypothetical example illustrated.

In fact many real sands show a tendency towards an inverse linear relationship between packing density and pore-fillers for values of less than 40–50 % of the latter, supporting the view that the mechanical effect of the material occupying the pores exerts an important influence on compaction.[113] The slope of the line varies between sandstones of one region, age and depth zone and another, presumably depending on palaeo-temperatures, nature of pore fluids, differences between overburden and pore pressure, and the effects of time. Further data need to be analysed in this way, but it appears from the foregoing that by extrapolating the best straight line fitting the points to meet the ordinate, one can read off the maximum packing density to be expected in a perfectly clean sandstone with no cement, buried under otherwise similar conditions to those samples on which the measurements were made.

Two points need stressing in connection with this method. First, the linear correlation between packing density and pore-fillers may be statistically less significant than often appears at first sight, as a result of the closure correlation[18] inherent in the fact that pore-fillers and porosity together add up to 100%. The second is that the pore space referred to is that visible in thin section; this is usually slightly less than the porosity measured by other methods, especially at low values, and the relationship between the two should be determined experimentally. In any case, only good quality thin sections prepared by impregnating the sandstone with coloured low-viscosity plastic should be used. The relationship between visible and porosimeter porosity values may depend on the colour of the dye used and the acuity of the microscopist's vision.

Finally, amongst the many factors influencing porosity reduction of all kinds in sandstones, there is the question of inhibition by hydrocarbons. Where oil or gas occupies the pores the relative permeability to water approaches zero. Transport of ions into or out of the sandstone is then limited to the very slow process of diffusion. Significant porosity reduction after the migration of petroleum into the reservoir would probably be limited to the margins of the accumulation, unless the process of solution transfer was effective. Several examples of the inhibiting effect of hydrocarbons have been reported. Füchtbauer[36] quotes examples of the influence of oil on secondary quartz cementation; Webb,[125] and Hancock and Taylor[49] demonstrate the control of clay mineral diagenesis by oil accumulation.

METHODS AVAILABLE FOR PREDICTION AND DELINEATION OF SANDSTONE RESERVOIRS

Looking for sandstone reservoirs in new petroleum provinces embraces three steps, usually with some overlap and feed-back: consideration of the regional geology on a basin-wide scale, reinterpretation and prediction of facies in the light of subsurface data from reconnaissance geophysics (especially seismic), and more detailed delineation and trend prediction following drilling.

The first of these steps involves all the established armoury of stratigraphic analysis, piecing together scattered evidence from libraries, outcrops, air photographs, and any boreholes or mines. Increasingly, however, expectations concerning the location, depositional environment, and even the petrology of sands hidden in the subsurface tend to be influenced by plate tectonic theories.

Plate Tectonics

The bearing of plate tectonics on petroleum exploration is dealt with in Chapter 1, but one application of this unifying concept can be touched on in this chapter. It can enable the geologist to predict the broad distribution and types of potential reservoir sands in untested areas, especially on the margins of continents (whether beneath a blanket of younger rocks or shelf seas, or both) by making use of relationships between tectonic style, location and timing of vertical movements, and the type of continental margin involved. The diversity of possibilities inherent in convergent or Pacific-type margins [27,115] is too great for elaboration here, but the simpler case of divergent or Atlantic-type margins illustrates the approach and has particular relevance to current exploration off the shores of north-west Europe, Greenland, and eastern North America.

According to King et al.,[63] the preliminary stages of development of divergent margins consist of thermal dilation of the mantle leading to regional uplift of the crust and dyke intrusion, followed by the formation of axial rifts with extensive block-faulting. When this occurs above sea level on an existing landmass, coarse continental clastics, partly derived from the uplifted edges of the rift by sub-aerial erosion, together with alluvium carried down the axis, can be expected to cover its floor; examples include the Triassic rifts of Western Europe, including the North Sea Viking Graben,[12] and possibly the Irish Sea and Cheshire Basins.[19] The reservoir value of the clastics will depend, as discussed elsewhere, partly on the nature of the rocks being eroded. If the rifting takes place below sea level, or is soon immersed beneath it, erosion of the rift flanks may be minimal, and the outward tilting attitude of the bounding fault blocks may prevent access of sandy sediment from distant lateral sources until the peripheral low areas have been filled, or until thermal contraction has lowered the flanks. Sand will then spill into the trough down a series of channels and deep-sea fans to spread as density flows of diminishing thickness and grain size across the bottom of the trough. Similar mechanisms may distribute sand lengthwise down the trough when either end is in communication with a suitable source—a condition particularly well met at so-called triple junctions when one arm is occupied by a major terrestrial drainage system terminating in a delta.[13] A variety of deep-sea sands therefore comes to occupy the floor of such sub-sea rifts, whereas the shelves on either side are occupied mainly by paralic and deltaic sands, although these may eventually also prograde across the rift-fill. Examples have been documented in the northern North Sea where these processes have created important reservoirs in the Jurassic and Tertiary.[11,81]

In the North Sea, tectonic development did not pass beyond the rifting stage. West of the British Isles rifting was followed by separation and the production of new ocean crust by intrusion along the axial zone. According to Roberts[90] this was terminated in the Rockall Trough after a separation of about 150 km 76 my ago, but further west the Atlantic Basin is still widening. The stratigraphy of these areas is known only from scattered JOIDES test holes and from seismic data, but according to the summary of King et al.[63] the continental shelf is likely to be built out beyond the original rift margin by a complex of varied deep, shallow marine, and even coastal plain sediments, whereas beyond it the continental rise appears to consist of a huge gently-dipping apron of coalescing turbidity deposit fans thinning slowly oceanwards.

Seismic Techniques

In the early stages of exploration, seismic work is concentrated on the problem of discovering structural traps. Later accumulation of subsurface knowledge, as a result of drilling, may enable seismic methods to play a significant part in delineating sandstone bodies. However, euphoric claims sometimes made for the direct recognition of lithology (as well as hydrocarbons) by reflection methods may unintentionally give a misleading impression. Seismologists do not attach quite the same meaning to the word 'lithology' as do sedimentologists. The 'bright spot' technique for detecting gas sands, discussed in Chapter 9, is a special case; the magnitude of the effect is sometimes so large that it is unlikely to have any alternative cause. In more general cases the information available from reflected energy is ambiguous, and 'determining lithology' in this context may mean no more than deciding which of two geologically plausible lithologies is the likelier.

Only two properties of seismic waves are at present available for distinguishing rock types—velocity and attenuation—and this imposes severe limitations on the extractable information. Interval velocities as routinely obtained have a relatively poor absolute accuracy, certainly not better than 2% (some practitioners reckon only on 10%) which deteriorates as the interval becomes thinner. According to Sheriff[103] the determination is usually impracticable for less than 200 ms TWT, which at a depth of 3 km is about 300 m thick. Only thick uniform formations with distinctive velocities can therefore be identified with certainty, using velocity alone. Consistent lateral variations may, however, suggest lithological changes or hydrocarbons in pores. The velocities of the potential reservoir rocks (sandstone, limestone and dolomite) depend far more on porosity than mineralogy. Thus, pure quartz sandstones range from 18 000 ft/s (5·5 km/s) at zero porosity down to about 10 500 ft/s (3·2 km/s) at 30 % porosity. Over the same porosity range limestones vary from 21 000 ft/s (6·4 km/s) to 11 000 ft/s (3·4 km/s), and dolomites from 23 000 ft/s (7·0 km/s) to 11 500 ft/s (3·5 km/s). Over-pressured shales show abnormally low velocities, and these zones may suggest directions of fluid migration, with effects discussed earlier under diagenesis.

Bed thickness can be resolved by recognition of the energy reflected from top and bottom only down to about $\frac{1}{4}$ wavelength, say 150–250 ft (46–76 m) at depths of 10 000–15 000 ft (3·1–4·6 km). Thinner sands give peaks of constant shape, but which vary in amplitude according to net sand thickness if porosity and hydrocarbon saturation do not change. Large changes in amplitude accompany changes in gas saturation ('bright spot'). The method has been successfully applied mainly in the Gulf Coast area. As lucidly explained by Dedman et al.[26] and Mateker and Chanseng,[75] the response can be investigated and calibrated using the synthetic seismogram. Sand isopachs can be drawn with a contour interval as small as 3 m.[70] Peak attenuation over thick intervals can similarly be utilised to determine sand/shale ratios, with a resolution of about 10 %. It will be appreciated that these methods work because the lithology can be assumed —or has been proved by adjacent drilling—to be simple, consisting in the cases quoted of comparatively uniform sands separated by thicker shales.

Most of the increase in velocity with depth which occurs in sandstones can be attributed to cementation and loss of porosity, rather than simply to compression.[37] In a geological province where thick masses of sand and shale can be expected to have accumulated under similar sedimentary conditions with no significant change in source, this relationship may allow cementation to be monitored.

In a more general way, inferences about environments of deposition can be obtained from consideration of the geometry of the strata drawn from seismic reflections. Resolution will still place limits to the fine detail that can be read, but larger features may give clues to smaller ones. The fore-setting relationships of lower Tertiary deltas in the North Sea have been impressively illustrated by Parker.[81] The geologist may be more alert to the significance of subtle changes in angular relationships, thickness, and reflection strength than the seismologist, anticipating where bed changes are likely to occur. It is necessary, however, for him to guard constantly against the temptation to look on seismic sections as geological cross-sections, rather than the complex interference figures they are. For guidance on the underlying principles and through the numerous pitfalls of interpretation the texts by Fitch,[33] and Tucker and Yorston[116] are particularly recommended.

Data from Boreholes
The third stage of prediction and delineation begins after drilling, whether or not this has proved the existence of a potential sandstone reservoir. It embraces measurement of reservoir parameters on core samples and with wireline logs, direct recognition of environments of deposition in cores, the imaginative interpretation of environments using cuttings and wireline logs, and finally the establishment of directional trends making use of dipmeter logs (and oriented cores if available), trend gradients between wells, and knowledge of the regional palaeo-slope.

Cores
There is no substitute for a conventional core, either for the measurement of physical properties or the recognition of environments, although side-wall slices taken after drilling can be helpful. Side-wall cores taken with explosive bullets may give unreliable porosity, permeability, and capillary pressure figures due to shattering and the introduction of drilling mud; they are difficult to interpret petrologically for the same reasons, but can yield information on sand grain size and mineralogy, cements other than clays, and provide some indication of bedding structures within the limitations of their small size.

Split longitudinally, continuous conventional cores give an unequalled view of the succession, easily recorded in compact permanent form by photography which can communicate some forms of data far better than a written description. The type, inclination, and scale of bedding features such as planar and trough cross-bedding, ripple-bedding, and graded-bedding, together with the wide variety of load casts, scour and tool marks, and post-depositional structures such as convolute-bedding, slump structures, and dish structures, can all be strongly indicative of particular environments.[82,87,96] They should not be read in isolation so much as considered in relation to the other types of structure present and their relative position in repeated *sequences*. The main limitations in the interpretation of sedimentary structures in conventional cores are imposed by size and uncertainties of orientation. Thus, it can be difficult to distinguish between planar-bedding and trough cross-bedding, unless the radius of the troughs is small, or to discriminate between the inclination due to foresetting, slump rotation, or regional dip. Cores are seldom oriented when cut; they can sometimes be oriented in the laboratory by matching the dip on shale bands to the structural dip, or by magnetic methods.

Despite the widespread use of micropalaeontology and palynology, macrofossils in cores can occasionally provide unique environmental evidence, for example from the standpoint of their orientation or method of preservation. Trace fossils (alternatively described as spreiten, lebenspuren, ichnofossils, burrows and bioturbation structures) are of far more general application, because they are widely distributed in sandstones as well as shales, and because their general form is independent of their geological age.[24,94] According to Seilacher[95] there is a general gradation from vertical burrows in shallow-water deposits to increasingly horizontal and more intricately patterned burrows in deep water. Lack of trace fossils in fine-grained deposits is a strong argument in favour of euxinic conditions, whilst their presence in any sediment is the strongest evidence against such conditions. Bioturbation can be an important homogenising agency, destroying horizontal laminations and enhancing vertical permeabilities. On the other hand, the same process can convert alternations of horizontally permeable sand with thin shale bands into uniform, argillaceous, non-permeable sand.

X-radiography has sometimes been used with success to detect bedding structures and bioturbation in apparently uniform sandstone.[56] Radiation from a tungsten filament source photographed on infrared film through a slab of sandstone a few millimetres thick is said by Rhoads and Stanley[89] to serve a similar purpose. A less exotic method is to flood a cleaned surface with cationic dye (e.g. a 1 % w/v solution of toluidine blue in water), rinse thoroughly, and examine whilst wet; the dye picks out differences in the content of clay minerals and organic matter not visible without treatment.

Careful comparison between laboratory-determined petro-physical properties and wireline log characteristics is an essential step in reservoir evaluation.[22] Use of helium is becoming common in Boyle's Law methods of determining porosity. It is also customary in deep wells to make an allowance for the compressibility of the sandstone; this is frequently responsible for a difference of several units of porosity between the surface and the reservoir.[22] Differences in permeability may be more significant.[127]

Ditch Samples

Only very small-scale sedimentary structures (fine laminations, small burrows) can be recognised in cuttings, but in the absence of cores these clues may be important.

Selley[97] has emphasised the significance of two key constituents of sandstones in the rapid assessment of the sedimentary environment. These are *glauconite*, the well-known indicator of marine deposition, and *carbonaceous matter*, which tends to be present in both marine and terrestrial deposits in inverse proportion to the amount of winnowing undergone. Presence/absence combinations define four classes of environment: glauconite plus carbonaceous matter—submarine channel and fan deposits, and turbidites; glauconite without carbonaceous matter —marine shelf sands, e.g. barriers, bars and shoals; carbonaceous matter without glauconite—fluvial, lacustrine and deltaic; neither carbonaceous matter nor glauconite—eolian. Like all simple methods, this one should be applied with circumspection.

The full range of diagnostic petrographic observations and measurements can be made on cuttings as readily as on core samples, provided they are not too finely comminuted (for example, by diamond bit or turbodrill), and as long as individual chips can be correlated via wireline logs with the formation drilled. Cuttings are cast into blocks in plastic and thin-sectioned in the usual way.

After a long lapse from favour there has been a revival of interest in the use of grain-size distributions in the interpretation of environments.[40,121] Cumulative frequency on a probability scale is plotted against size on a logarithmic scale. In general, sands plot as combinations of two or three straight-line segments; it is thought by proponents of the method that these can be equated with the traction load, saltation load and suspension load of the depositing medium. The slope of the segments (sorting of the respective populations) and the positions of their intersections are believed to be diagnostic, and published examples show a strong empirical relationship between certain shapes of curve and certain environments, summarised in Fig. 1. Many more observations are needed to put the method on a sound basis; there is a particular shortage of examples of various marine shelf environments. In certain cases restriction of supply of particular grain sizes, imposed by the source, may prevent the 'ideal' distribution from developing, but this influence sometimes has time–stratigraphic value.

Wireline Logs

Great improvements have been made in the scope and quality of wireline logs. Combinations of suitable logging tools can, in principle, determine rock composition completely in terms of the lithology and fluid content on the basis of one measurement per independent variable.[15,34] In practice combinations in common use, including gamma ray or self potential, together with one or more 'porosity' logs (sonic, neutron, or bulk density), plus resistivity logs covering deep and shallow penetration, often give satisfactory results. In some cases (the North Sea Jurassic is one) problems arise because of the unusual abundance of dense minerals such as pyrite and micas, and radioactive minerals such as micas and zircon,[55,91] or because of abnormal water salinities. In straightforward cases a variety of computer programs is available for converting the readings directly into porosity, producible hydrocarbons, residual hydrocarbons, and water saturation, together with measures of clay and carbonate content. The last two determinations may be of assistance in studies of reservoir cementation, but it will be noted that there is no method of distinguishing between primary and secondary quartz, so questions concerning quartz cementation and pressure solution cannot be solved. However, once a particular interval has become familiar to the geologist in several wells, preferably by the study of cores, but alternatively by repeated observations of cuttings in conjunction with wireline logs, surprisingly close interpretation of variations in cementation can be made by rapid visual comparison of, say, a gamma ray, sonic and resistivity log.

Qualitative interpretation of grain size from self potential or gamma ray logs has probably been widely used within the industry for some time. The method has been described by Pirson,[84] Jageler and Matuszak,[60] and Selley;[97] further examples have been given recently by Conybeare.[23] It relies on the generalisation that in most cases the coarser and cleaner a sand, the lower is its rate of gamma ray emission and the more negative its self potential. Either log can, therefore, be used to distinguish between sand units in which the grain size remains substantially constant throughout their thickness, from those in which it increases upwards or downwards; indentations in the curves also discriminate between uniform sands and those with thin shaly streaks. Since the direction of coarsening is a specific attribute of sands laid down under certain conditions, this simple method of observation provides an extremely powerful tool in the preliminary diagnosis of the environment of deposition. Some examples are given in Fig. 1; it should be noted that the vertical scale, as well as the log character need to be considered. Where possible it should be backed up by other methods. Problems may arise through the difficulty in correctly picking the bases of sedimentation units, through units being too thin for log resolution, or as a result of unsuspected mineralogical changes. The self-potential log in particular can be influenced by many factors other than grain size, including hydrocarbon content and the character of adjacent beds, whilst statistical variation on the gamma ray record can mask lithological subtleties.

Dipmeter Logs

Apart from the use of dipmeter logs in helping to orientate cores and recognise faults and unconformities, an extensive methodology has been developed for their environmental interpretation.[14,38,84] Basically, these methods depend on the recognition of patterns of ascending or descending increase in dip, constant dip, and randomly varying dip directions, which are characteristic of various sand body geometries (Fig. 1). There is little doubt that in suitable cases the technique can be helpful. The main drawbacks are associated with poor dip detection—many sands are too uniform to give reliable internal dip measurement by resistivity methods— and also the standard minimum resolution of 2 m is too coarse to disentangle dips in thin units.

CONCLUSION

Petroleum geology is passing through that uncomfortable transition in the development of any science—from descriptive to quantitative. It is, perhaps, particularly uncomfortable for the geologist, because the process has advanced further in reservoir engineering and geophysics than in geology. This raises problems of communication between members of a team in which the fullest co-operation is vital if the more elusive reservoirs are to be found without deteriorating success ratios; it is not made easier by the fact that the different disciplines attract people with dissimilar mental characteristics.

To the mathematically trained petroleum engineer a methodology which takes factual data such as porosity at a small number of wells, relates it to a hypothetical depositional model such as a delta, attempts to infer from minimal data what the unseen subsurface form of that delta is, and then transforms this chimera into a spatial prediction of porosity, is at best suspect. Since there are so many uncertainties, he may ask, why not stick to the measured parameters of direct interest and make statistical predictions about subsurface variations using recognised numerical methods? This attitude overlooks the fact that few, if any, geological data are isotropic.

Some geometric arrangements of properties are more likely than others, and it is the claim of sedimentology that these patterns can be recognised and their broad outlines predicted with reasonable assurance. The role of the geologist, to use the apt words of Moore,[77] is to introduce informed bias into the interpretation.

The difficulty in applying this philosophy is knowing when to stop. The geologist who tries to squeeze his material to the limit of the predictive clues it contains, runs the risk of being labelled academic. Usually, the pace of petroleum exploration determines the cut-off point, and many a study becomes a post-mortem in circumstances when it might be argued that if enough detail could have been assimilated and interpreted in less time, it might have provided a viable guide to subsequent drilling.

Can sedimentological studies be made in sufficient depth to be meaningful, yet fast enough to make a genuine contribution to exploration? Environmental interpretation is not too badly placed in this respect; once the diagnostic criteria are thoroughly understood, their acquisition from logs, cuttings and cores is not particularly time-consuming for an experienced investigator. The detailed elucidation of diagenetic history is a different matter. Various forms of analysis—porosity, permeability, capillary pressure, X-ray mineral determination, etc.—have been mechanised and speeded up, but the vital relationships between one rock component and another remain essentially problems for the petrological or electron microscope.

Partial mechanisation can already lighten and accelerate petrological work and help to reduce some of its subjectivity. The preparation of high-quality thin sections can now be automated. The point-counting microscope stage has long been available to facilitate quantitative determinations of composition and grain size; it can now be plugged into a calculator and plotter which will, for instance, draw grain-size curves directly.[107] Primary and secondary quartz can, in favourable circumstances, be distinguished from one another by cathodo-luminescence; when bombarded with electrons in an evacuated chamber they emit light of different colours depending on the trace elements present.[106] Secondary X-ray emission can also map the distribution of different elements.[71] Descriptive, as well as measured, data can be converted to numerical form, and computer software is available to store, manipulate, retrieve and finally plot the results on maps and cross-sections,[50] but no satisfactory alternative to the geologist yet exists between the microscope and the computer. In medical science, the success that has been achieved in the automation of routine examination of histological preparations under the microscope, using an image scanning system and computerised pattern-recognition methods, indicates a possible way ahead. The problems should not be under-estimated, however. For instance, there are the number of variables to be dealt with; some twenty common components of environmental and diagenetic significance are likely to occur in sandstones and need to be recognised and recorded. Particle shape, colour, and spatial relationships could account for at least another twenty data entries, whilst size distributions involve recording a similar number of classes. Discrimination poses greater problems. Many decisions required in compiling the data listed above have a high subjective content. For example: is this area of clay allogenic or authigenic? What was the original shape and size of that quartz grain which now has an authigenic overgrowth? The clues needed to answer these and similar questions may be extremely subtle; the sedimentary petrologist draws on years of experience as well as reasoning to take all the relevant detail into account in making his decisions. When the complexity and sophistication that would need to be designed into a machine to take his place, and the cost of its development are considered, the time taken by the geologist to make a detailed analysis is put into truer perspective.

Yet, a start has been made. Conley and Davis,[21] and Conley[20] have described automatic petrographic analyses of simple limestones and sandstones using electronic image analysis. The machine can be taught to recognise shapes and textures. Each class of constituent can be identified, grains counted, areas measured, and statistics derived for immediate data reduction or display.

Perhaps the future role of the sedimentary petrologist will be as an interpreter and an arbiter of cases too complex for the machine to decide unaided. Rather than the machine analysis of rock properties and mineralogy in the laboratory, however, there is much to be said for putting the analysis underground, enabling continuous records to be obtained. The present day computerised lithological wireline log has many limitations— but how many would have believed that it would have been possible at all twenty years ago?

ACKNOWLEDGEMENT

The author is grateful to V. C. Illing & Partners for encouragement and the provision of time and facilities for the preparation of this chapter.

REFERENCES

1. ALLEN, J. R. L. (1965). A review of the origin and characteristics of Recent alluvial sediments, *Sedimentology*, **5**, 89–191.
2. ALLEN, J. R. L. (1969). Notes towards a theory of concentration of solids in natural sands, *Geol. Mag.*, **106**, 309–21.
3. ALLEN, J. R. L. (1970). Studies in fluviatile sedimentation: a comparison of fining-upwards cyclothems, with special reference to coarse-member composition and interpretation, *J. Sediment. Petrol.*, **40**, 298–323.
4. ALMON, W. R., FULLERTON, L. B. and DAVIES, D. K. (1976). Pore space reduction in Cretaceous sandstones through chemical precipitation of clay minerals, *J. Sediment. Petrol.*, **26**, 89–96.
5. ARNTSON, R. H. (1963). Effect of temperature and confining pressure on the solubility of calcite at constant CO_2 concentrations, *Abstracts, Geol. Soc. Am. Ann. Meet.*, 40.
6. ATWATER, G. I. and MILLER, E. E. (1965). The effect of decrease in porosity with depth on future development of oil and gas reserves in South Louisiana, *Am. Assoc. Petrol. Geologists Ann. Convention*, New Orleans.
7. BEARD, D. C. and WEYL, P. K. (1973). Influence of texture on porosity and permeability of unconsolidated sand, *Am. Assoc. Petrol. Geologists Bull.*, **57**, 349–69.
8. BLANCHE, J. B. (1973). The Rotliegendes sandstone formation of the United Kingdom sector of the southern North Sea basin, *Trans. Inst. Min. Met.*, B, **82**, 85–9.
9. BOER, R. B. DE (1977). On the thermodynamics of pressure solution— interaction between chemical and mechanical forces, *Geochim. Cosmochim. Acta*, **41**, 249–56.
10. BOER, R. B. DE, NAGTEGAAL, P. J. C. and DUYVIS, E. M. (1977). Pressure-solution experiments on quartz sand. *Geochim. Cosmochim. Acta*, **41**, 257–64.
11. BOWEN, J. M. (1975). The Brent oilfield, in: *Petroleum and the Continental Shelf of North-west Europe*, 1 (A. W. Woodland, ed.) Applied Science Publishers, Barking, 353–62.
12. BRENNAND, T. P. (1975). The Triassic of the North Sea, in: *Petroleum and the Continental Shelf of North-west Europe*, 1 (A. W. Woodland, ed.) Applied Science Publishers, Barking, 295–312.
13. BURKE, K. and DEWEY, J. F. (1973). Plume-generated triple junctions: key indicators in applying plate tectonics to old rocks, *J. Geol.*, **81**, 406–33.
14. CAMPBELL, R. L. (1968). Stratigraphic applications of dipmeter data in Mid-Continent, *Am. Assoc. Petrol. Geologists Bull.*, **52**, 1700–19.
15. CAMPBELL, R. L. (1972). Recent advances in log evaluation, *World Petroleum*, **43**, 22–30.
16. CANT, D. J. and WALKER, R. G. (1976). Development of a braided-fluvial facies model for the Devonian Battery Point Sandstone, Quebec, *Can. J. Earth Sci.*, **13**, 102–19.
17. CARRIGY, M. A. and MELLON, G. B. (1964). Authigenic clay mineral cements in Cretaceous and Tertiary sandstones of Alberta, *J. Sediment. Petrol.*, **34**, 461–72.

18. CHAYES, F. (1971). *Ratio Correlation*, Univ. Chicago Press, Chicago and London.
19. COLTER, V. S. and BARR, K. W. (1975). Recent developments in the geology of the Irish Sea and Cheshire Basin, in: *Petroleum and the Continental Shelf of North-west Europe*, 1 (A. W. Woodland, ed.), Applied Science Publishers, Barking, 61–76.
20. CONLEY, C. D. (1976). Variation in petrographic characteristics of reservoir sandstones determined by image analysis, *Abstracts, Geol. Soc. Am.*, **8**, 819.
21. CONLEY, C. D. and DAVIS, J. C. (1973). Carbonate petrography by pattern recognition, *Am. Assoc. Petrol Geologists Bull.*, **57**, 399–406.
22. CONNER, D. C. and KELLAND, D. G. (1974). Piper field, UK North Sea interpretive log analysis and geologic factors, *Trans. 3rd Europ. Formation Eval. Symp. Lond.*, A.
23. CONYBEARE, C. E. B. (1976). Geomorphology of oil and gas fields in sandstone bodies, *Developments in Petroleum Science*, **4**, Elsevier, Amsterdam, Oxford, New York.
24. CRIMES, T. P. and HARPER, J. C. (eds.) (1970). *Trace Fossils*, Seel House Press, Liverpool.
25. CURRAY, J. R. (1956). Dimensional grain orientation studies of Recent coastal sands, *Am. Assoc. Petrol. Geologists Bull.*, **40**, 2440–56.
26. DEDMAN, E. V., LINDSEY, J. P. and SCHRAMM, M. W. (1975). Stratigraphic modelling: a step beyond bright spot, *World Oil*, **180**(6), 61–5.
27. DEWEY, J. F. and BURKE, K. (1974). Hot spots and continental break-up: implications for collisional orogeny, *Geology*, **2**, 57–60.
28. DOEGLAS, D. J. (1962). The structure of sedimentary deposits of braided rivers, *Sedimentology*, **1**, 167–90.
29. ERNST, W. G. and BLATT, H. (1964). Experimental study of quartz overgrowths and synthetic quartzites, *J. Geol.*, **72**, 461–9.
30. EYNON, G. and WALKER, R. G. (1974). Facies relationships in Pleistocene outwash gravels, southern Ontario: a model for bar growth in braided rivers, *Sedimentology*, **21**, 43–70.
31. FATT, I. (1958). Pore structure in sandstones by compressible sphere-pack models, *Am. Assoc. Petrol. Geologists Bull.*, **42**, 1914–23.
32. FISHER, W. L. and BROWN, L. F. (1969). Delta systems and oil and gas occurrence, in: *Delta systems in the exploration for oil and gas* (Fisher *et al.*, eds.) Res. Colloq., Bur. ec. Geol., Univ. Texas.
33. FITCH, A. A. (1976). *Seismic Reflection Interpretation*, Gebrüder Borntraeger, Berlin, Stuttgart.
34. FORD, M. E., BAINS, A. J. and TARRON, R. D. (1974). Log analysis by linear programming—an application to the exploration for salt cavity storage locations, *Trans. 3rd Europ. Formation Eval. Symp. Lond.*, B.
35. FOTHERGILL, C. A. (1955). The cementation of oil reservoir sands and its origins, *Proc. 4th World Petroleum Congress*, 301–14.
36. FÜCHTBAUER, H. (1974). Some problems of diagenesis in sandstones, *Bull. Centre Rech. Pau-SNPA*, **8**, 391–403.
37. GARDNER, G. H. F., GARDNER, L. W. and GREGORY, A. R. (1974). Formation velocity and density—the diagnostic basics for stratigraphic traps, *Geophysics*, **39**, 770–80.

38. GILREATH, J. A. and MARICELLI, J. J. (1964). Detailed stratigraphic control through dip computations. *Am. Assoc. Petrol. Geologists Bull.*, **48**, 1902–10.

39. GINSBURG, R. N. (1975). *Tidal Deposits*, Springer-Verlag, Berlin, Heidelberg, New York.

40. GLAISTER, R. P. and NELSON, H. W. (1974). Grain-size distributions: an aid in facies identification, *Can. Petrol. Geol. Bull.*, **22**, 203–40.

41. GLENNIE, K. W. (1970). Desert sedimentary environments, *Developments in Sedimentology*, **14**, Elsevier, Barking, Essex.

42. GLENNIE, K. W. (1972). Permian Rotliegendes of north-west Europe interpreted in light of modern desert sedimentation studies. *Am. Assoc. Petrol Geologists Bull.*, **56**, 1048–71.

43. GLENNIE, K. W. (1977). Diagenesis of Permian Rotliegendes sandstones in Leman Bank and Sole Pit areas of UK southern North Sea, *J. Geol. Soc.*, in press.

44. GRATON, L. C. and FRASER, H. J. (1935). Systematic packing of spheres, *J. Geol.*, **43**, 785–909.

45. GROULT, J., REISO, L. H. and MONTADERT, L. (1966). Reservoir inhomogeneities deduced from outcrop observations and production logging, *J. Petrol. Tech.*, **18**, 883–91.

46. HALBOUTY, M. T. (1972). Rationale for deliberate pursuit of stratigraphic, unconformity and paleogeomorphic traps, *Am. Assoc. Petrol Geologists Bull.*, **56**, 537–41.

47. HANCOCK, J. M. and SCHOLLE, P. A. (1974). Chalk of the North Sea, in *Petroleum and the Continental Shelf of North-west Europe*, **1** (A. W. Woodland, ed.) Applied Science Publishers, Barking, 413–28.

48. HANCOCK, N. J. (1977). Possible causes of Rotliegend sandstone diagenesis, *J. Geol. Soc. Lond.*, **135**, in press.

49. HANCOCK, N. J. and TAYLOR, A. M. (1977). Clay mineral diagenesis and oil migration in the Middle Jurassic Brent Sand, *J. Geol. Soc. Lond.*, **135**, in press.

50. HARBAUGH, J. W. and MERRIAM, D. F. (1968). *Computer Applications in Stratigraphic Analysis*, Wiley, New York.

51. HEALD, M. T. and ANDEREGG, R. C. (1960). Differential cementation in the Tuscarora Sandstone, *J. Sediment. Petrol.*, **30**, 568–77.

52. HEALD, M. T. and LARESE, R. E. (1974). Influence of coatings on quartz cementation, *J. Sediment. Petrol.*, **44**, 1269–74.

53. HOBDAY, D. K. and READING, H. G. (1972). Fair weather versus storm processes in shallow marine sand bar sequences in the Late Precambrian of Finmark, North Norway. *J. Sediment. Petrol.*, **42**, 318–24.

54. HOBSON, G. D. and TIRATSOO, E. N. (1975). *Introduction to Petroleum Geology*, Scientific Press Ltd, Beaconsfield, England.

55. HODSON, G. M., FERTL, H. and HAMMACK, G. W. (1975). New log data give better North Sea well completions, *World Oil*, **181**(4), 60–5.

56. HOWARD, J. D. (1968). X-ray radiography for examination of burrowing in sediments by marine invertebrate organisms, *Sedimentology*, **11**, 249–58.

57. HOWER, J., ESLINGER, E. V., HOWER, M. E. and PERRY, E. A. (1976). Mechanism of burial metamorphism of argillaceous sediment: 1. Mineralogical and chemical evidence, *Bull. Geol. Soc. Am.*, **87**, 725–37.

58. HRABAR, S. V. and POTTER, P. E. (1969). Lower West Baden (Mississippian) sandstone body of Owen and Greene counties, Indiana, *Am. Assoc. Petrol. Geologists Bull.*, **53**, 2150–60.

59. HSU, K. J. (1977). Studies of Ventura field, California, II: lithology, compaction, and permeability of sands, *Am. Assoc. Petrol. Geologists Bull.*, **61**, 169–91.

60. JAGELER, A. H. and MATUSZAK, D. R. (1972). Use of well logs and dipmeters in stratigraphic-trap exploration, in: *Stratigraphic Oil and Gas Fields* (R. E. King, ed.) Am. Assoc. Petrol. Geologists Mem. 16, SEG Spec. Publ. 10, 107–35.

61. JOHNSON, H. D. (1977). Shallow marine sand bar sequences: an example from the late Precambrian of North Norway. *Sedimentology*, **24**, 245–70.

62. KAHN, J. S. (1956). The analysis and distribution of the properties of packing in sand-size sediments. Part 1. On the measurement of packing in sandstones, *J. Geol.*, **64**, 385–95.

63. KING, L. H., HYNDMAN, R. D. and KEEN, C. E. (1975). Geological development of the continental margin of Atlantic Canada, *Geosci. Can.*, **2**, 26–35.

64. KING, R. E. (ed.) (1972). *Stratigraphic Oil and Gas Fields—Classification, Exploration Methods, and Case Histories*, Am. Assoc. Petrol. Geologists Mem. 16, SEG Spec. Publ. 10.

65. KLEMME, D. (1977). Giant fields contain less than 1 % of world's fields but 75 % of reserves, *Oil Gas J.*, **75**(10) 164.

66. KRUMBEIN, W. C. and MONK, G. D. (1942). Permeability as a function of the size parameters of unconsolidated sand, *Petrol. Tech.*, AIME. Tech. Publ. 1492, 5, 1–11.

67. KUMAR, N. and SANDERS, J. E. (1974). Inlet sequence: a vertical succession of sedimentary structures and textures created by the lateral migration of tidal inlets, *Sedimentology*, **21**, 491–532.

68. LATTMAN, L. H. (1973). Calcium carbonate cementation of alluvial fans in southern Nevada, *Bull. Geol. Soc. Am.*, **84**, 3013–28.

69. LERBEKMO, J. F. (1957). Authigenic montmorillonoid cement in andesitic sandstones of Central California, *J. Sediment. Petrol.*, **27**, 298–305.

70. LINDSEY, J. P., SCHRAMM, M. W. and NEMETH, L. K. (1976). New seismic technology can guide field development, *World Oil*, **182**(7), 59–63.

71. LONG, J. V. P. (1963). Recent advances in electron-probe analysis, in: *Advances in X-ray Analysis*, **6**, Plenum Press, New York, 276–90.

72. MACKENZIE, D. B. (1972). Primary stratigraphic traps in sandstone, in: *Stratigraphic Oil and Gas Fields* (R. E. King, ed.) Am. Assoc. Petrol. Geologists Mem. 16, SEG Spec. Publ. 10, 47–63.

73. MARIE, J. P. P. (1975). Rotliegend stratigraphy and diagenesis, in: *Petroleum and the Continental Shelf of North-west Europe*, **1** (A. W. Woodland, ed.) Applied Science Publishers, Barking, 205–12.

74. MAST, R. F. and POTTER, P. E. (1963). Sedimentary structures, sand shape fabrics, and permeability, II. *J. Geol.*, **71**, 548–65.

75. MATEKER, E. J. W. and CHANSHENG, W. U. (1971). True amplitude data can indicate lithology, *World Oil*, **172**(5) 73–5.

76. MAXWELL, J. C. (1964). Influence of depth, temperature, and geologic age on

porosity of quartzose sandstone, *Am. Assoc. Petrol. Geologists Bull.*, **48,** 697–709.

77. MOORE, P. G. (1967). The use of geological models in prospecting for stratigraphic traps, *Trans. 7th World Petrol. Cong.*, 481–5.

78. MORGRIDGE, D. L. and SMITH, W. B. (1972). Geology and discovery of Prudhoe Bay field, Eastern Arctic Slope, Alaska, in: *Stratigraphic Oil and Gas Fields* (R. E. King, ed.) Am. Assoc. Petrol. Geologists Mem. 16, SEG Spec. Publ. 10, 489–501.

79. MORROW, N. R. (1971). Small-scale packing heterogeneities in porous sedimentary rocks, *Am. Assoc. Petrol. Geologists Bull.*, **55,** 514–22.

80. OOMKENS, E. (1970). Depositional sequences and sand distribution in the postglacial Rhône Delta complex, in: *Deltaic Sedimentation—Modern and Ancient* (J. P. Morgan, ed.) SEPM. Spec. Publ., 15, 198–212.

81. PARKER, J. R. (1975). Lower Tertiary sand development in the Central North Sea, in: *Petroleum and the Continental Shelf of North-west Europe*, 1 (A. W. Woodland, ed.) Applied Science Publishers, Barking, 447–54.

82. PETTIJOHN, F. J., POTTER, P. E. and SIEVER, R. (1973). *Sand and Sandstones*, Springer-Verlag, Berlin, Heidelberg, New York.

83. PIRSON, S. J. (1963). *Handbook of Well Log Analysis—for Oil and Gas Formation Evaluation*, Prentice-Hall Inc., Englewood Cliffs, New Jersey.

84. PIRSON, S. J. (1970). *Geologic Well Log Analysis*, Gulf Publ. Corp., Houston.

85. PITTMAN, E. D. (1972). Diagenesis of quartz in sandstones as revealed by scanning electron microscopy, *J. Sediment. Petrol.*, **42,** 507–19.

86. POLASEK, T. L. and HUTCHINSON, C. A. (1967). Characterization of non-uniformities within a sandstone reservoir from a fluid mechanics standpoint, *Proc. 7th World Petroleum Congress*, 2, Elsevier, Barking, Amsterdam, New York. 397–408.

87. REINECK, H. E. and SINGH, I. B. (1973). *Depositional Sedimentary Environments*, Springer-Verlag, Berlin, Heidelberg, New York.

88. RENTON, J. J., HEALD, M. T. and CECIL, C. B. (1969). Experimental investigation of pressure solution of quartz, *J. Sediment. Petrol.*, **39,** 1107–17.

89. RHOADS, D. C. and STANLEY, D. J. (1966). Transmitted infrared radiation: a simple method for studying sedimentary structures, *J. Sediment. Petrol.*, **36,** 1144–9.

90. ROBERTS, D. G. (1975). Tectonic evolution of the Rockall Plateau and Trough, in: *Petroleum and the Continental Shelf of North-west Europe*, 1 (A. W. Woodland, ed.), Applied Science Publishers, Barking, 77–92.

91. ROBERTS, H. V. and CAMPBELL, R. L. (1974). The application of Coriband to the micaceous Jurassic sandstones of the northern North Sea Basin, *Trans. 3rd Europ. Formation Eval. Symp.*, London, D.

92. RUST, B. R. (1972). Structure and process in a braided river, *Sedimentology*, **18,** 221–46.

93. SCHMIDT, V. and MCDONALD, D. A. (1977). The role of secondary porosity in the course of sandstone diagenesis, *CSPG Reservoir*, **4,** 1.

94. SEILACHER, A. (1964). Biogenic sedimentary structures, in: *Approaches to Paleoecology* (J. Imbrie and N. Newell, eds.), John Wiley, New York, London, Sydney, 296–316.

95. SEILACHER, A. (1967). Bathymetry of trace fossils, *Marine Geology*, **5,** 413–28.

96. SELLEY, R. C. (1976). *An Introduction to Sedimentology*, Academic Press, London, New York, San Francisco.

97. SELLEY, R. C. (1976). Subsurface environmental analysis of North Sea sediments, *Am. Assoc. Petrol. Geologists Bull.*, **60**, 184–95.

98. SELLEY, R. C. (1977). Porosity gradients in North Sea oilbearing sands, *J. Geol. Soc. Lond.*, **135**, in press.

99. SHARP, W. E. and KENNEDY, G. C. (1965). The system $CaO–CO_2–H_2O$ in the two-phase region calcite plus aqueous solution, *J. Geol.*, **73**, 391–403.

100. SHEARMAN, D. J., *et al.* (1977). Sandstone diagenesis, *J. Geol. Soc.Lond.*, **135**, in press.

101. SHELTON, J. W. (1964). Authigenic kaolinite in sandstone, *J. Sediment. Petrol.*, **34**, 102–111.

102. SHELTON, J. W. (1973). Models of sand and sandstone deposits: a methodology for determining sand genesis and trend, *Oklahoma Geol. Surv. Bull.*, 118.

103. SHERIFF, R. E. (1976). Inferring stratigraphy from seismic data, *Am. Assoc. Petrol. Geologists Bull.*, **60**, 528–42.

104. SHINN, E. A., HALLEY, R. B., HUDSON, J. H. and LIDZ, B. H. (1977). Limestone compaction: an enigma, *Geology*, **5**, 21–4.

105. SIEVER, R. (1959). Petrology and geochemistry of silica cementation in some Pennsylvanian sandstones, in: *Silica in Sediments* (H. A. Ireland, ed.) SEPM. Spec. Publ., 7, 55–79.

106. SIPPEL, R. E. (1968). Sandstone petrology, evidence from luminescence petrography, *J. Sediment. Petrol.*, **38**, 530–54.

107. SLATT, R. M. and PRESS, D. E. (1976). Computer program for presentation of grain-size data by the graphic method, *Sedimentology*, **23**, 121–31.

108. SMITH, D. B. (1970). The palaeogeography of the British Zechstein, *3rd Symp. on Salt*, North Ohio. Geol. Soc. Am. Publ., 20–3.

109. SPRUNT, E. S. (1976). Reduction of porosity by pressure solution: experimental verification, *Geology*, **4**, 463–6.

110. STALDER, P. J. (1973). Influence of crystallographic habit and aggregate structure of authigenic clay minerals on sandstone permeability, *Geologie en Mijnbouw*, **52**, 217–20.

111. STUBBLEFIELD, W. L., LAVELL, J. W., SWIFT, D. J. P. and McKINNEY, T. F. (1975). Sediment response to the present hydraulic regime on the Central New Jersey Shelf, *J. Sediment. Petrol.*, **45**, 337–58.

112. SWIFT, D. J. P., DUANE, D. B. and ORRIN, H. P. (1973). *Shelf Sediment Transport: Process and Pattern*, John Wiley, Chichester.

113. TAYLOR, J. C. M. (1977). Control of diagenesis by depositional environment within a fluvial sandstone sequence in the Northern North Sea Basin, *J. Geol. Soc. Lond.*, **135**, in press.

114. THOMAS, J. B. (1977). Thermodynamic considerations of diagenesis in argillaceous sandstone, *J. Geol. Soc. Lond.*, **135**, in press.

115. THOMPSON, T. L. (1976). Plate tectonics in oil and gas exploration of continental margins, *Am. Assoc. Petrol Geologists Bull.*, **60**, 1463–1501.

116. TUCKER, P. M. and YORSTON, H. J. (1973). Pitfalls in seismic interpretation, *Soc. Exp. Geoph.*, Monograph Series, 2.

117. VAN VEEN, F. R. (1975). Geology of the Leman gas-field, in: *Petroleum and the*

Continental Shelf of North-west Europe, **1** (A. W. Woodland, ed.) Applied Science Publishers, 223–32.

118. VERRIEN, J. P., COURAND, G. and Montadert, L. (1967). Applications of production geology methods to reservoir characteristics analysis from outcrops observations, *Proc. 7th World Petroleum Congress*, Elsevier, Barking, Amsterdam, New York, 425–46.

119. VISHER, G. S. (1965). Use of vertical profile in environmental reconstruction,, *Am. Assoc. Petrol. Geologists Bull.*, **49**, 41–61.

120. VISHER, G. S. (1965). Fluvial processes as interpreted from ancient and Recent fluvial deposits, in: *Primary Sedimentary Structures and their Hydrodynamic Interpretation* (G. V. Middleton, ed.) SEPM Spec. Publ., 12, 116–32.

121. VISHER, G. S. (1969). Grain size distribution and depositional processes, *J. Sediment. Petrol.*, **39**, 1074–1106.

122. WALKER, T. R. and CRONE, A. J. (1974). Mechanically-infiltrated clay matrix in desert alluvium, *Abstract Geol. Soc. Am. Ann. Meet.*, **9**, 998.

123. WAUGH, B. (1970). Formation of quartz overgrowths in the Penrith Sandstones (Lower Permian) of north-west England as revealed by scanning electron microscopy, *Sedimentology*, **14**, 309–20.

124. WAUGH, B. (1970). Petrology, provenance and silica diagenesis of the Penrith Sandstone (Lower Permian) of north-west England, *J. Sedim. Petrol.*, **40**, 1226–40.

125. WEBB, J. E. (1974). Relation of oil migration to secondary clay cementation, Cretaceous sandstones, Wyoming, *Am. Assoc. Petrol. Geologists Bull.*, **58**, 2245–9.

126. WILLIAMS, J. J., CONNER, D. C. and PETERSON, K. E. (1975). The Piper oil field U.K. North Sea: a fault block structure with Upper Jurassic beach/bar reservoir sands, in: *Petroleum and the Continental Shelf of North-west Europe*, **1** (A. W. Woodland, ed.) Applied Science Publishers, Barking, 363–78.

127. ZOBACK, M. D. and BYERLEE, J. D. (1975). Permeability and effective stress, *Am. Assoc. Petrol. Geologists Bull.*, **59**, 154–8.

Chapter 6

DELTAIC FACIES AND PETROLEUM GEOLOGY

R. C. SELLEY

Royal School of Mines, Imperial College, London, UK

SUMMARY

Deltas often contain large quantities of oil and gas because the deltaic process injects porous reservoir sands far out into marine basins with abundant source beds.

Deltas contain a multiplicity of potential reservoirs whose distributions are related to the fluvial, tidal and wave processes which operate upon them. Deltas contain many different types of trap, not only structural traps of tectonic origin, but also growth faults and roll-over anticlines, as well as diverse stratigraphic traps.

The insulating effect of rapidly deposited over-pressured deltaic shales with low thermal conductivity allows the build-up of abnormally high temperatures and pressures which favour large-scale hydrocarbon generation.

Deltaic deposits may be identified and correlated in the subsurface from seismic data and wireline logs, as well as from conventional core studies.

INTRODUCTION

A very large proportion of the world's sedimentary cover is composed of ancient deltaic deposits. Petroleum explorationists have long known that ancient deltas are often areas where substantial amounts of oil and gas have been generated and trapped.

Modern and ancient deltas have been extensively described in the geological literature. This chapter begins by briefly reviewing the deltaic

197

process and modern and ancient deltas. Then follows a discussion of the criteria by which the petroleum explorationist may recognise ancient deltas from subsurface data, and, more importantly, how this knowledge may be used to predict the distribution of porous reservoir beds in deltaic facies.

There is a diffuse literature describing hydrocarbon accumulations in ancient deltas, but few published accounts analyse why deltas are such significant hydrocarbon provinces, or where productive fairways may be present. The chapter concludes with a review of these two topics.

THE DELTAIC PROCESS AND MODERN AND ANCIENT DELTAS

To begin this account of deltas and petroleum, it is necessary to have a clear understanding of the basic physical processes which operate in a delta. Consider a standing body of water at whose margin there is an aperture at the water level emitting a sediment-laden flood (Fig. 1). The current velocity will diminish radially from the aperture; thus, as sediment settles out, the coarsest particles will be deposited nearest the aperture and the silt and clay will be deposited in quieter water some distance away. Hence, sediment particle size will decrease radially from the aperture.

As sedimentation continues the deposits adjacent to the aperture will build up, layer by layer, to the water level. These layers will then prograde out into the body of water, giving rise to a series of dipping foreset beds. Migration of the foreset beds causes coarser sediment to overlie finer. In this way a delta forms an upward-coarsening sequence of grain size.

At the top of the deltaic sediment prism, however, the current velocity maintains a scoured distributary channel through the topset part of the delta. Sediment is deposited on either side of this channel to form levees which may build up to the water level. Fine-grained clay may settle out in the sheltered bays on the outer sides of the levees. As the process continues the distributary channel and its flanking levees build out so far that sediment begins to choke the channel's opening. A levee crevasses near to the source, and a new levee-flanked channel builds out in a different direction from the first, and so it continues.

This simplified view of the deltaic processes suggests that a delta should show the following features:

1.   An upward-coarsening profile of grain size ranging from the deep-water clays of the bottomset to the silts and sands of the delta slope (foreset).

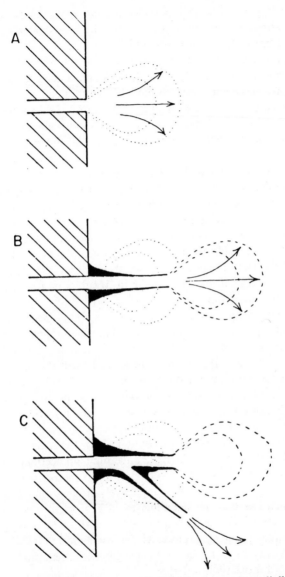

Fɪɢ. 1. Stages of development in an ideal delta system. A: radially-decreasing velocities from jet mouth deposit concentric arcs of sand, silt and clay. B: delta progrades forcing a channel through marginal levees. C: channel mouth chokes, levee ruptures and a new delta builds out from the crevasse. Reproduced by permission of Academic Press.

2.   An upper surface (topset) consisting of a series of radiating distributary channels, often infilled by sands, which cut through fine-grained silts and clays of the levees and inter-distributary bays.

3.   In vertical profile, sequences of the type described above should be repeated cyclically. In three-dimensions, a delta should consist of many sediment lobes within which sediment grain size should increase upwards and towards the source. This repetition reflects the fact that with their habit of crevassing levees and of switching distributary channels, deltas contain built-in cycle-generators.

If we now transfer this model from the realms of hypothesis to the margin of a sea, we can also consider other features that may be expected to be present in an ancient delta. Considering first the biota, it is reasonable to suppose that the pro-delta muds will contain marine fossils, although often with derived fossils of continental origin. The topset of each deltaic sequence, however, will contain largely continental and brackish remains. Because of the large area of the topset at, or very close to, sea level, the water-logged sediment may provide an ideal substrate for extensive plant growth, with concomitant peat formation.

Another aspect to consider is the fate of an abandoned delta lobe after the sediment supply has moved elsewhere. The weight of overburden causes the muds near the base of the lobe to compact. The surface of an abandoned delta lobe thus slowly subsides beneath the sea.

In a sheltered situation the topset swamps may be immediately overlain by open-sea muds. In exposed situations, however, the transgression may be marked by strand-line sand-bodies where the sea reworks the top of the abandoned delta, winnowing out the fines and carrying them seawards.

Let us move now from the realms of hypothesis to the real world, and see how this ideal delta model which we have built up accords with reality.

To start with, it is interesting to compare this picture with the modern Mississippi delta, one of the most intensively studied modern deltas, and the one which has been most publicised in geological literature. As will shortly be shown, it is extremely unfortunate that the Mississippi delta has received so much publicity, because it is, in fact, a most atypical delta. None the less, it forms a natural link between the ideal delta model previously discussed, and more typical deltas.

Figure 2 is a map showing the main sedimentological features of the Mississippi delta. More details will be found in papers by Fisk,[7] Coleman and Gagliano,[2] Kolb and van Lopik,[15] and Gould.[9]

Figure 2 shows that the modern Mississippi delta agrees well with our

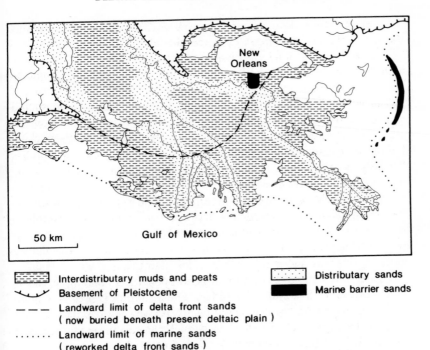

Interdistributary muds and peats
Basement of Pleistocene
Landward limit of delta front sands
(now buried beneath present deltaic plain)
Landward limit of marine sands
(reworked delta front sands)

Distributary sands
Marine barrier sands

FIG. 2.   Distribution of major sand facies in the modern Mississippi delta (a fluvially-dominated delta). Reproduced by permission of Academic Press.

previously proposed model. A radiating system of linear sands cut through extensive inter-distributary lagoon and bay muds, and swamp peats. The origin of the radiating sands is quite complex, however. Sand has not been deposited as much within the distributary channels, as on bars at the seaward mouths of levee systems. Nevertheless, the geometry of the potential reservoir sands is much the same.

Detailed coring, dating and subsurface mapping have defined numerous sub-delta lobes of Quaternary to Recent age. The cyclic sedimentary pattern thus formed, however, is due, not only to the crevassing of levees, but is intimately linked with glacially-related sea-level changes. The formation of upward-coarsening cycles is well demonstrated, albeit on a relatively small scale, around the present active mouth of the delta. Repeated surveying of the Head of Passes has shown the genesis, advance and abandonment of several crevasse-splay sub-deltas.[2]

Abandonment and subsidence of a major sub-delta lobe is well seen to the east of New Orleans, where the arcuate Chandeleur barrier islands formed as the sea transgressed the abandoned St Bernard delta.

This review of the modern Mississippi delta shows that it conforms well with the ideal delta model, although its evolution has been complicated by glacial sea-level changes.

The Mississippi delta is very well-known for two reasons. First, because upstream, at least, is an area of extensive settlement and land use; and secondly, because the Mississippi delta system was initiated early in the Cenozoic. These earlier deposits contain a major hydrocarbon province beneath the present-day delta. It has, therefore, been studied to aid the understanding of its antecedents.

The modern Mississippi delta is anomalous, and a global view shows that it is very different from most deltas. There are two main reasons for this. In the first instance, the sediment load of the Mississippi is largely clay and silt, with only about 10% of very fine sand. Most deltas, ancient and modern, have far higher sand/shale ratios than this.

The second anomaly is that the Mississippi delta is building out into a sheltered sea with a low tidal range. Thus, marine processes are too ineffective to rework the deltaic sediments. As we shall see, there are few modern deltas in which distributary channels can extend very far seawards without being truncated by marine shore-face sands.

The radiating 'birdfoot' geometry of the Mississippi delta is seldom to be found, except in the non-tidal sheltered waters of lakes.

Let us now look at some more typical deltas where marine processes are intimately associated with delta building. A useful distinction can be made at this point between marine-dominated deltas where wave action is important, and marine-dominated deltas where tidal currents play the dominant role in sediment redistribution.

Examples of the first case are shown by the deltas of the Nile (Fig. 3) and the Niger. These are totally different in their geometry, and hence in the distribution of reservoir sands, from the Mississippi delta. In wave-dominated deltas such as those of the Nile and the Niger, sediment is no sooner deposited at the mouth of a distributary than it is reworked by wave action. Fine-grained clay and silt are winnowed out and settle out from suspension in the deeper water of the pro-delta. Sand, on the other hand, is thrown back on to the edge of the delta in an arcuate fringe of shore-face sands. These may form barrier islands, capped even by aeolian dunes right around the delta periphery. In such deltas the inter-distributary areas will still be areas where clay is deposited and peat may form (mangrove swamps in the case of the Niger), but this will tend to occur in freshwater lacustrine environments, rather than in the marine bays and brackish tidal lagoons of fluvially-dominated deltas, like the Mississippi.

Because of the inhibiting effect of the arcuate barrier sands, delta switching is constrained, and is limited to minor crevasse-splay formation. Cyclicity in wave-dominated deltas is more likely to be related to external causes, such as sea-level changes, rather than to major delta switching.

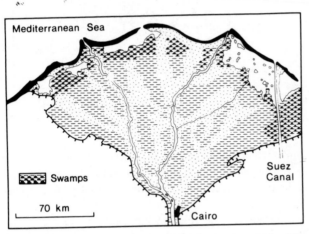

FIG. 3.   Distribution of major sand facies in the modern Nile delta (a wave-dominated delta). Reproduced by permission of Academic Press.

Finally, it is necessary to consider the second type of marine-dominated delta, in which tidal currents play the major role in redistributing deltaic sediment.

Several tidal deltas have been studied and described in south-east Asia, such as the Mekong, the Klang, Langat and Ganges–Brahmaputra deltas.[3,19] Figure 4 illustrates the last-named.

In these areas of large tidal range, powerful tidal currents scour up and down the deltaic distributaries. The channels, instead of showing a radiating pattern, occur in sub-parallel braided systems. If the wave energy is weak coastal barrier sands do not form where the delta meets the sea. Instead, a broad flat area of swamps passes transitionally through tidal flats into an open marine environment.

This review of the delta process and of modern deltas is necessarily brief. The main point of the analysis is to show how the process controls the environment and hence the facies. The three different types of delta process

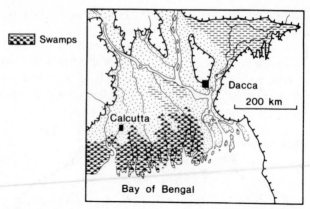

FIG. 4.   Distribution of major sand facies in the modern Ganges–Brahmaputra delta (a tide-dominated delta). Reproduced by permission of Academic Press.

deposit sands which may be potential hydrocarbon reservoirs. Early recognition of the processes which dominated the deposition of an ancient delta may help to predict the location, geometry and orientation of potential reservoirs.

Table 1 summarises the relationship between deltaic processes and potential hydrocarbon reservoirs.

TABLE 1

THE RELATIONSHIP BETWEEN DELTAIC PROCESSES AND POTENTIAL RESERVOIR TYPE

Dominant process	Environment	Examples	Reservoir types
Fluvial	Radiating distributaries and intervening bay muds and marsh pans	Mississippi	Radiating channel and mouth bar sands
Marine Tides	Tidal flats and swamps cut by braided tidal channels	Ganges–Brahmaputra	Sub-parallel tidal channels
Waves	Distributaries truncated by barrier islands	Nile Niger	Arcuate shore-face sands

RECOGNITION OF SUBSURFACE DELTAIC DEPOSITS

Because deltas host many oil and gas fields, much attention has been paid to the criteria by which deltaic facies may be recognised from subsurface data. This has been greatly helped by knowledge of modern deltas and deltaic processes, and by studies of ancient deltas which crop out at the surface.[6,20,29]

This section of the chapter reviews the criteria by which a delta may be identified in the subsurface. This will be done by examining, one by one, the five parameters which define a sedimentary facies, viz. geometry, lithology, sedimentary structures, palaeo-currents and palaeontology.

Recognition of Deltaic Geometry

In petroleum geology, traditionally it is the geometry of a facies, or of an individual reservoir, that is the last criterion to be established. The object of the whole exercise is, in fact, to establish the geometry of rock units. There are numerous elegant published maps of radiating deltaic distributary channel sand systems, but it is often germane to ask at what stage of development the deltaic origin of the sands was established, and what part such a map played in the location of boreholes.

There is, however, one particular line of approach by which the deltaic origin of sedimentary facies may be established prior to drilling. Large-scale progradational foresetting can often be seen on seismic cross-sections. This, of itself, demonstrates the progradational nature of the sediments in question, and, after allowing for compactional effects, gives some indication of the depth of water into which the delta built (an esoteric detail of relatively small economic significance).

A classic example of this is the Palaeocene delta of the North Sea, in which deep-sea sands deposited on submarine fans at the foot of the delta slope house the Forties and other oil accumulations.[22] Careful seismic mapping can, therefore, define the overall geometry of a delta before it has ever been drilled.

As will be shown, there are certain parts of a deltaic prism which are more likely to contain hydrocarbons than others. In particular there are optimum sand/shale ratios where the right balance of reservoir to seal favours hydrocarbon entrapment.

Here again, geophysics may be helpful in defining the optimum fairways. This is done by careful study of interval velocities which may be used to delineate sand and shale units, and hence permit the construction of tentative sand/shale ratio maps prior to drilling. For many companies this

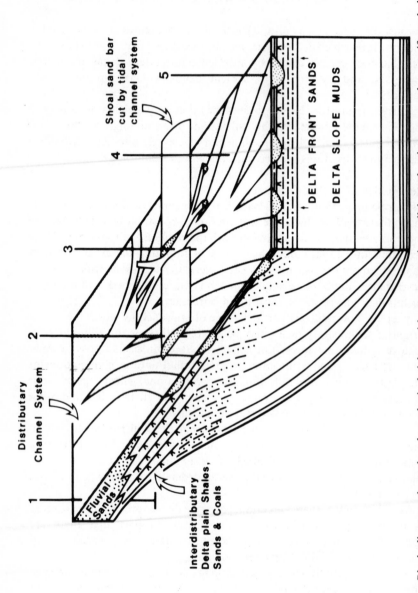

Fig. 5. Block diagram through a hypothetical delta, with accompanying well logs showing gamma log motifs and the

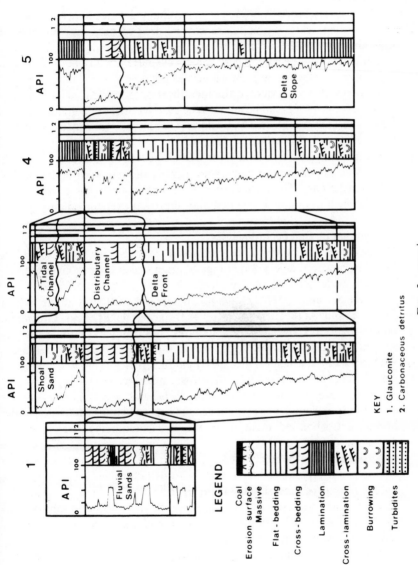

Fig. 5—contd.

technique is still in the experimental stage, and there are various problems such as, for example, where an over-pressured shale has the same velocity as a porous sand.

This is a very broad approach to determining deltaic facies distribution. On a smaller scale, seismic work can be used to map the geometry of individual sand bodies within a delta.[11,17] This is generally most effective at shallow depths, especially where some wells have already been drilled so that interval velocities are known accurately.

To conclude this review of the identification of ancient deltas by their geometry, two points can be made. When the geometry of deltaic facies has been accurately established by drilling, it is generally too late to apply this knowledge in hydrocarbon exploration.

Seismic mapping can define the overall geometry of a delta, produce tentative sand/shale ratio maps, and identify discrete sand bodies within deltaic sediments. These techniques are rudimentary at the present time, but they will doubtless be refined and improved over the coming years.

Recognition of Deltaic Lithologies

Deltas are, by their very nature, composed of terrigenous clastic sediment. There is, however, a very wide range of grain size in their sediments and a similarly wide range of sand/shale ratios.

On the one hand is the modern Mississippi river with only about 10% of very fine sand. At the other extreme are the gravel deltas of east Greenland and of the Red Sea, where fanglomerates pass from land into deep water with little change in slope, grain size or any other sedimentary feature.

One rock type which is particularly characteristic of (but not unique to) deltas, is coal. Because deltas often form extensive water-logged coastal plains they are areas where plants may thrive and be preserved as beds of peat. Many ancient deltas contain beds of coal or lignite, often with rootlet horizons. Carbonised plant fragments are commonly disseminated in deltaic sands and clays.

In the same way that there is no unique deltaic lithology, neither are there characteristic types of sediment texture. A delta involves a wide variety of sedimentary processes, ranging from traction currents in fluvial channels, through beach processes at the shore-face, to grain-flow and turbidites at the foot of some delta slopes. There is, therefore, no particular type of sediment texture which characterises a delta.

It will be shown that, although there are no unique lithologies or textures which characterise deltas, there are, nevertheless, characteristic lithological sequences which may be used to identify deltas in the subsurface.

Recognition of Deltaic Sedimentary Structures

Although lithologies can be determined and identified from well cuttings and geophysical logs, sedimentary structures can only be studied in those rare instances where a well has been cored. Even so, there is the same problem as was encountered when discussing deltaic lithology and texture. Many sedimentary processes operate within a deltaic environment. Sedimentary structures are controlled by the process rather than by the environment *per se*. Thus, the whole range of sedimentary structures may occur in ancient deltaic facies. On the topset cross-bedded sand channels will be cut into cross-laminated flaser-bedded clays and silts, often intensively burrowed. Desiccation cracks and rootlet horizons may testify to sub-aerial exposure.

At the periphery of the delta, flat-bedding may characterise the shore-face and barrier sands, often with burrowing in their lower part.

The silts and clays of the delta slope are generally laminated, and may show evidence of slumping and mud diapirism, often on a vast scale.

As already mentioned, some deltas have aprons of deep-sea sands at the foot of their slopes. Here the whole range of sedimentary structures which typify grain-flow and turbidite processes can be found.

Thus, because the processes which operate on deltas are so diverse, the sedimentary structures are similarly varied.

Subsurface Recognition of Diagnostic Deltaic Sequences

The preceding sections show that the lithologies and sedimentary structures of deltas are diverse and intrinsically non-diagnostic. Sequences of lithology and structures, however, are frequently diagnostic and, more important still, can be determined from subsurface data.

One of the most important lessons to be learnt from the study of modern sediments is that certain environments have characteristic sequences of grain size and sedimentary structures, e.g. the upward-fining sequence of a fluvial channel and the upward-coarsening sequence of a barrier island.

It has already been discussed how the deltaic process forms a major upward-coarsening profile from the pro-delta mud to the top of the delta slope. The topset of a delta is characterised by various grain-size profiles, generally on a smaller scale than that of the delta slope. The topset may show upward-fining sequences in the distributary channels and upward-coarsening sequences in the crevasse-splay and shore-face sands.

These grain-size profiles are extremely useful in the recognition and analysis of subsurface environments. It has long been known that the gamma and spontaneous potential (SP) logs often reflect grain-size profiles in boreholes.

Now, as already discussed, no grain-size profile is specific to any one environment. Upward-fining channels may occur in a wide range of environments, from submarine canyon to alluvial fan. Thus, no gamma or SP log motif is diagnostic of an environment, deltaic or otherwise. Log profiles can aid environmental interpretation, however, when coupled with the presence or absence of glauconite and carbonaceous detritus. This technique has been described in detail elsewhere, and will only be summarised here.[25-8]

Glauconite is a mineral which can be recognised in well samples. It forms within a marine environment during early diagenesis just below the sediment/water interface, whence it may be winnowed and come to rest in any marine environment from beach to submarine fan. Glauconite is very susceptible to weathering and is virtually unknown as a second-cycle mineral. For these reasons glauconite in a sand indicates its marine origin, although the converse is not true.

Carbonaceous detritus (coal, lignite and kerogen) can also be identified in well samples. Its presence indicates deposition in a poorly-winnowed environment, tending to be anaerobic and with no oxidation. Carbonaceous detritus is generally absent from well-winnowed high-energy environments.

When combined, information on the presence or absence of glauconite and carbonaceous detritus can be used to characterise various environments. High-energy marine sands (beaches, barrier islands, shoals and tidal channels) tend to contain glauconite and to be devoid of plant matter. Fluvial and deltaic distributary channels, crevasse-splays and delta slopes tend to contain carbonaceous matter and to lack glauconite. Deep-sea sands may contain both.

The presence or absence of these constituents can thus help to identify environments when combined with gamma and SP log motifs (grain-size profiles). For example, an upward-coarsening log motif can be indicative of a prograding barrier sand, or of a crevasse-splay. One would expect to find glauconite present in the former, and carbonaceous detritus in the latter.

Figure 5 shows the various log responses to be expected in a series of wells drilled through a delta, and the way in which the distribution of glauconite and carbonaceous detritus differentiates sand bodies with similar grain-size profiles.

This technique is useful in paralic sands in general, and in deltaic sands in particular, because of the characteristic abundance of carbonaceous material of plant origin within deltas. As a general rule, carbonaceous detritus will be most abundant in fluvial-dominated deltas, while the ratio of glauconite to carbonaceous detritus will increase in wave- and tidally-influenced marine-dominated deltas.

Ideally, deltas should be identified in the subsurface from the careful study of sedimentary sequences in lengthy cores. Where this material is not available, however, then the technique of combining log character with the vertical distribution of glauconite and carbonaceous detritus may be very useful.

Figure 6 shows a typical facies analysis of a well in the North Sea, made using this technique aided by a study of a cored interval.

Recognition of Deltaic Palaeo-current (Dipmeter) Patterns

Ancient deltaic sediments contain stratigraphically trapped oil which may not be found by structural mapping. Facies analysis is vital in searching for stratigraphic traps and palaeo-current analysis is employed as part of this approach. The latter is possible from boreholes by using the dipmeter log which shows the amount and direction of dip of the strata.

Accounts of how this technique works will be found in the Schlumberger and Dresser-Atlas manuals, and an introduction to the principles of stratigraphic dipmeter interpretation has been given by Campbell[1] and McDaniel.[18] Specific accounts of dipmeter patterns in deltaic facies have been presented by Gilreath and Stephens.[8]

Basically two types of dipmeter motif occur in deltaic sands (Fig. 7): one characterises the channel, the other the progradational slope.

In a channel major accretionary foresets are often formed on point-bars dipping into the channel axis, i.e. their strike is coincident with current direction. As the channel is infilled the amount of dip decreases. Upward-decreasing dip patterns of constant dip direction may often be seen in channel sands (as identified from log character and sedimentary structures when cores are available).

Additional information may be extracted where the channel is thick and the dip computations are closely spaced. Major point-bar foresets are commonly composed of smaller-scale (<1 m) foresets which dip down-current. These may sometimes be seen as dips of between 15–25°, which are perpendicular to the major foresets, indicating the actual current direction (Fig. 7).

Thus, where an oil accumulation has been found within a channel, it is possible to determine the trend of the reservoir and the direction towards

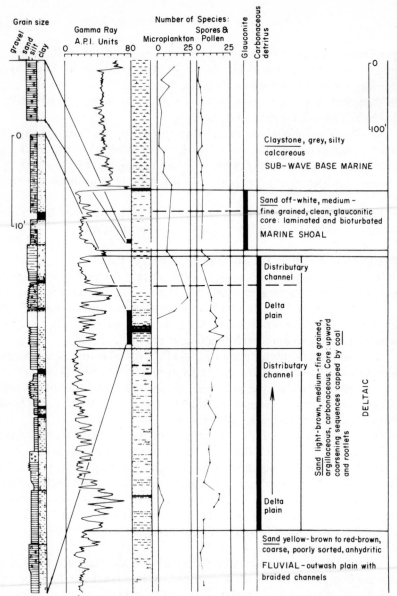

FIG. 6. Facies analysis of a North Sea well log. This shows a marine transgression in which continental red beds are ultimately buried beneath marine shale. Two prograding increments of carbonaceous deltaic sediments are overlain by glauconitic shoal sands which were deposited as the sea reworked the abandoned delta. Reproduced by permission of the American Association of Petroleum Geologists.[27]

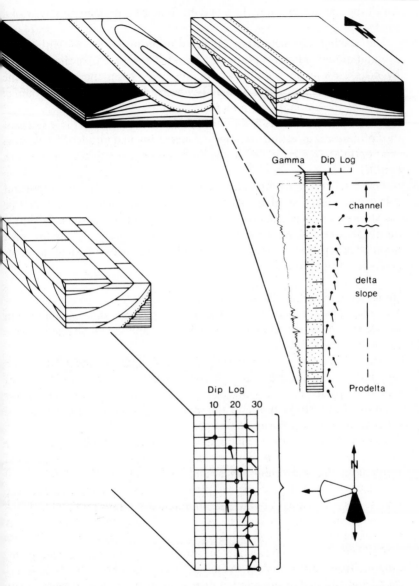

FIG. 7.   Block diagram showing dipmeter motifs in deltaic sand bodies. Major channel foresets dip west into channel axes. minor foresets (shown inset) dip south in the direction of delta progradation, shown also by southerly depositional dip in delta slope.

which the channel axis lies. This may be done by a dipmeter analysis of a single well.

The second characteristic deltaic dip motif is found within the progradational sediments of the delta front. These sands are characterised by major low-angle bedding which dips away from the mouth of the distributary channel from which they were deposited. As the channel progrades the amount of dip increases upward.

Hence, the characteristic dip motif for a delta-front sand is a series of dips of uniform direction and upward-increasing magnitude. The channel (and optimum reservoir) lies in the direction opposite to the dip azimuth (Fig. 7).

This brief review of dipmeter interpretation shows what an important role palaeo-current analysis plays in unravelling deltaic facies. It cannot be too strongly emphasised, however, that, for a variety of reasons, the simple motifs discussed above are not always present; secondly, the dip log must never be interpreted on its own, but must be integrated with all of the other geological information.

Recognition of Deltaic Biota

Two points have already been made concerning the fossils encountered in deltaic facies: the abundance of plant detritus, and the range from marine to non-marine fossils.

Deltas build out vast expanses of water-logged sediment which are favourable areas for plant growth and, for the same reason, dead plant material is preserved in this anaerobic, stagnant environment. The cypress swamps of the Mississippi and the mangrove swamps of the Niger and other tropical deltas illustrate this point.

Ancient deltaic facies are characterised, therefore, by a high percentage of coal and lignite beds, and by carbonaceous detritus disseminated throughout the intervening sands and shales. The Carboniferous Coal Measures of Northern Europe and the Pennsylvanian coal province of the Illinois basin are cases in point. The abundance of humic material in deltas is also important, because it is a significant source for gaseous hydrocarbons, as will be shown later.

Coal beds form in several different sub-environments on the delta plain, ranging from drifted plant material, which comes to rest in inter-distributary bays and abandoned distributary channels, and from in situ coals overlying root beds and palaeo-sols (seat-earths) which succeed levees and crevasse-splays. As a general rule, extensive plant formation and preservation characterise the phase of abandonment of a delta lobe. Coal beds, therefore, tend to overlie the shallow topset part of a delta and to be overlain by marine shales deposited during the ensuing advance of the sea.

For this reason coal beds are the most reliable marker horizons to use when correlating deltaic facies. Not only are they far more extensive laterally than individual sand and shale beds, but they show up easily in well cuttings and on geophysical logs.

Turning now to the marine and brackish biota of deltas, it is obvious that these characterise the lower and distal parts of deltaic facies.

The brackish waters of inter-distributary bays and lagoons contain micro-fossils, such as ostracods, and macro-fossils such as lamellibranchs. Oyster biostromes are forming, for example, in the bays of the modern Mississippi delta. Extensive bioturbation may also occur in these environments.

In the pro-delta environment an abundant marine fauna may be expected. This is sometimes the case, especially where deltas are building out on to a carbonate shelf environment. Microplankton and foraminifera may be present in the pro-delta and delta-slope shales. It has often been observed, however, that some distal deltaic shales are completely barren of marine biota. This reflects the fact that although the delta slope and pro-delta are, physiographically, marine environments, the sediment deposited there is wholly of continental origin. Since it is settling out of freshwater there is often no mixing with marine water containing potential marine micro-fossils.

THE PETROLEUM GEOLOGY OF DELTAIC FACIES

It is well known that ancient deltaic facies contain large reserves of hydrocarbons. Coal has already been mentioned but oil and gas are frequently present in deltaic sands. The hydrocarbons in deltas tend to have a high ratio of gas to oil, and the oils are often heavy but of low sulphur content.[14] It is not possible to give a realistic estimate of the amount of the world's oil and gas reserves trapped in deltaic facies, because of the problems of definition. Is one to consider the up-dip alluvial plain, and the deep-sea sands at the foot of some deltas?

This chapter concludes with an examination of the reasons why deltas are such important hosts for hydrocarbons, and consideration of the spatial distribution of hydrocarbons within a deltaic prism.

Five conditions must be fulfilled for an accumulation of oil or gas to form and to be preserved. There must be a source bed, a reservoir bed, and a seal or cap rock. These three must be arranged in such a way that hydrocarbons can be trapped. Finally, the source bed must have been heated sufficiently to have generated hydrocarbons. These five conditions will now be discussed within the context of deltaic sedimentation.

Hydrocarbon Source Beds in Deltaic Facies

First, then, let us consider deltaic source beds. The reasons for the abundant formation and preservation of plant material have already been discussed. It has been observed that there is a general relationship between the nature of the organic matter in the source bed and the type of hydrocarbons which it has generated. It is particularly significant that, when the percentage of hydrogen in the kerogen of the source bed is less than six the source bed tends to generate gas, whereas when it is more than six it tends to generate oil.[16]

The humic material of deltas tends to have less than 6 % of hydrogen, and for this reason deltas often generate hydrocarbon gases. (We are leaving aside here the biogenic generation of methane which takes place at shallow depths and low temperatures. This is of no economic significance.)

Deltas are not only sources of gas, however. The transgressive marine shales which envelop deltaic lobes are often source rocks, and, lacking humic material, tend to generate oil. One can see, therefore, that deltaic facies contain abundant source beds; gas-prone sources within each delta lobe, and oil-prone sources in the marine shale envelopes.

It is interesting to note that hydrocarbons may migrate out of the deltaic sediments themselves to be trapped in overlying non-deltaic reservoirs. A classic example of this occurs in the southern North Sea basin where gas has been driven off from the Carboniferous Coal Measures and has been trapped in the Rotliegendes and younger Permian, and in the Triassic cover.[23] There is a correlation between the distribution of the resultant gas fields and the stratigraphy and thermal history of the underlying deltaic source beds.[5]

Distribution of Reservoirs in Deltaic Facies

Deltas contain abundant reservoir sands, although, as already mentioned, the ratio sand/shale varies from delta to delta, as well as vertically and laterally within a single delta. Indeed, the most important aspect of the deltaic process is that it is a mechanism for transporting sand into a marine sedimentary basin.

Considered in more detail, there are a number of particular environments within a delta which are favourable for sand sedimentation.

First, there are the channel systems to consider, ranging from the fluvial channels of the alluvial plain, through the distributary channels of the delta itself, to the tidal channels of marine-dominated deltas.

There are two important points to note about channels, however. Channels are conduits for sediment transport, rather than sites of sand deposition. Thus, it is not uncommon to find that many channels are ultimately infilled by mud after they have been abandoned by a major crevassing episode. To locate a channel is not, therefore, automatically to locate a sandstone reservoir.

The second point to note is the lateral frequency of channels across the delta plain. This is likely to be a function of the sand/clay ratio of the sediment load, and of the length of duration of one particular sea level. Obviously, the higher the ratio of sand to mud brought into the delta, the higher will be the proportion of sand to shale. In ancient deltaic facies it may be possible to map closure on the upper surface of a delta lobe, but it may not be possible to guarantee that a channel sand will be present at a specific well location. In some instances, however, the sea level remained constant for a long period of time. Thus, as the distributary channels switch to and fro, there comes a time when the topset of the delta plain consists of a sheet of porous sand which is made up of a multitude of coalesced channels. The so-called 'Massive Sand' of the Jurassic Brent delta of the North Sea is a case in point. Another example is the Lafourche 'horsetail' sub-delta of the Mississippi. In such instances it is relatively easy to locate porosity in these laterally extensive sheet sands.

Moving from the distributary channels to the delta front, a different set of problems are encountered. Delta-front sands and, to a lesser extent, crevasse-splay sands tend to be of much greater lateral extent than channel sands. They are, therefore, easier to find and, once found, to correlate. The petrophysical properties are often different, however, since porosity and permeability tend to increase with increasing grain size and decreasing clay content. The best porosities and permeabilities tend to occur in the coarser winnowed sands of the channels. Lower porosities and permeabilities characterise the finer argillaceous sands of crevasse-splays and the delta fronts. This illustrates a common problem of petroleum exploration; that the lateral continuity of a reservoir varies inversely with its porosity and permeability.

This same generalisation holds true for the marine sand bodies associated with deltas. Extremely high porosities may be found in the well-winnowed sands which occur at the periphery of wave-dominated deltas. Unlike channels, however, barrier beaches are environments of sand deposition, not just of transportation. Like channels, though, they may be of limited lateral extent.

Concluding this review of reservoirs in deltas, it can be seen that there are various environments in which porous sands can be deposited. There is a close relationship between environment and the petrophysics, geometry and trend of any particular sand body. The overall relationship between process and reservoirs in deltaic facies has been previously pointed out (Table 1).

Cap Rocks in Deltaic Facies

The third parameter needed for an oil field to occur is the cap rock or seal. This will now be considered in the context of deltaic facies. A general principle of petroleum exploration has been derived from the Gulf Coast of North America. This is that the most productive part of a delta is the 'break-up' zone where the sand/shale ratio is about 50:50, as continental sands inter-finger down-dip with marine shales. The rationale behind this concept is that down-dip of this zone reservoir sands are too few, while up-dip of this zone there are insufficient shales to form adequate seals to inhibit hydrocarbon migration.

At the down-dip limit of the break-up zone the problem is to locate the reservoir, but at the up-dip limit of the break-up zone the problem is to locate the seal. A specific example of this has been documented from south-east Texas by Gregory.[10] He has shown how a particular sand of great lateral extent is only productive to seaward of the up-dip limit of the *Clavulina byramensis* marine shale which lies directly above it.

On a more regional scale, Dailly[4] has introduced the concept of the 'pendulum effect'. The weight of a deltaic pile of sediment may be so great as to depress the Earth's crust, thus forming a moat around the depocentre. Major re-entrants on either side of a delta lobe are sites of extensive fine-grained sedimentation in bay and tidal flat environments. After some time a delta will switch into one of the peripheral moats, burying the laterally extensive clay unit. Through time the delta switches to and fro like a pendulum, burying a number of these moat sheet shales.

This is an interesting concept which Dailly developed in the Niger delta, and he speculates that the prolific oil zones in this hydrocarbon province underlie moat shales, which are regionally significant cap rocks.

Hydrocarbon Traps in Deltaic Facies

A trap is the fourth parameter required for an oil field to need discussion within the deltaic context. Deltas are favourable areas for hydrocarbon entrapment, because they contain both structural and stratigraphic traps.

For a large delta to form, an initial crustal depression is essential, although, as already noted, once it is established the weight of deltaic sediment may be sufficient to depress the crust.

The initiation of a depression is, of itself, a sign of tectonic instability. Thus, deltaic facies generally overlie an unstable basement which is frequently a site of crustal tension. This is manifest by block-faulting. The resultant horsts and grabens may play a role in controlling early deltaic palaeo-geography and facies. Of more importance in this context, however, is the way in which the overlying deltaic beds may be arched and faulted along earlier lines of structural weakness. Thus, hydrocarbons may be trapped in anticlines draped over deeply-buried horsts in the pre-deltaic basement. Similarly, basement faults may be rejuvenated, leading to the formation of fault traps in the overlying deltaic cover.

This situation is well seen in the molasse trough of northern Italy, where gas in younger Tertiary deltaic sediments is trapped in structures related to earlier tectonic lineaments in the basement.

Not all structural traps in deltas are due to basement-related tectonics. Deltas also contain their own trap-making processes. When large amounts of sediment are rapidly deposited, as in a delta, the normal rate of de-watering and compaction is inhibited by rapid burial. When this happens the clays in the lower part of the prism become abnormally pressured and remain extremely plastic. Differential loading of the topset cover begins to take place, leading to the formation of sub-vertical faults near the surface which decrease in dip downwards until apparently they become coincident with bedding plane shear surfaces deep down in the pro-delta. Stratigraphic units thicken basinwards across such faults, and the amount of throw increases with depth. Such faults are termed growth faults.

Growth faults are particularly important because of the way in which they can act as hydrocarbon traps. The first point of importance is that not only do stratigraphic units thicken on the down-thrown side of the fault, but the sand/shale ratio increases as well. The percentage of reservoir section increases, therefore, on the down-thrown side of the fault. This is due, presumably, to sand being trapped in the hollow which occurs on the down-thrown side of the fault where it approaches the sea-floor.

Oil and gas may be trapped near to growth faults in two ways. The trap may be a classic fault trap where the sand abuts against the fault. Alternatively, oil may be trapped in a roll-over anticline where the beds turn over to dip into the plane of the fault (Fig. 8).

The Niger delta is the classic example of a hydrocarbon province with many growth fault and roll-over traps.[30] Growth faults also occur in the Gulf Coast Tertiary delta province, but the great continuation of some of these (such as the Vicksburg Flexure) argues for a deeper-seated basement control.

Under-compacted clays in deltas are also responsible for another type of trapping mechanism. Mud lumps, or shale diapirs, make their way up

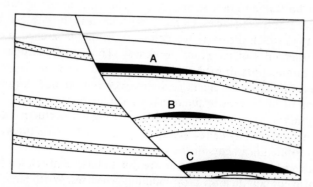

FIG. 8.   Cross-section to show hydrocarbon traps associated with growth faults and roll-over anticlines. Note how sediments thicken and are sandier on the downthrown side of the fault. A: Fault trap; B and C: Roll-over anticlines.

towards the surface due to differential loading of the topset part of the delta. These structures may dome up the overlying sediments, forming anticlinal traps and peripheral fault traps.

Within the Gulf Coast province there are many fields associated with diapiric movements, but in this case the diapirs are generally of salt, not clay, which has moved up from the pre-deltaic Jurassic Louann series.

Finally, deltas may contain a diversity of stratigraphic traps. These range from the most proximal to the most distal parts of a delta. At its up-dip edge traps may be present where fluvial sands wedge out between impermeable pre-deltaic basement and transgressive marine shales. Within the delta plain oil may be stratigraphically trapped within the distributary channels and the mouth-bar sands. In those deltas with aprons of deep-sea sands there is a potential fairway for hydrocarbon entrapment at the up-dip pinch-out of the submarine channel sands.

Concluding this review of hydrocarbon entrapment deltas, it can be seen that there are a multiplicity of possible traps. These range from basement-related structural traps, through the quasi-structural/stratigraphic growth fault and diapiric traps, to a diversity of stratigraphic traps.

Geothermal Gradients in Deltaic Hydrocarbon Provinces

The last of the five parameters to consider is the temperature required to generate oil. It is now widely held that a temperature of over 150°F is needed for major oil generation, while at temperatures of 350–400°F oil tends to break down to hydrocarbon gases.[24]

Because, as has already been indicated, deltas form in areas of crustal instability, they occur in areas of relatively high heat flow. In terms of basin type, the deltas which are major hydrocarbon provinces tend to occur in continental margin basins. Intra-cratonic basins tend to be dominated by continental facies; marine and deltaic deposits are relatively rare.

Continental margin basins have relatively high geothermal gradients and, where a delta has prograded over the continental margin on to oceanic crust, the gradient may be very high indeed. The Niger delta and the Gulf Coast of North America are the two areas whose heat flows are known in greatest detail.

Data from the Gulf Coast show a gradual decrease in geothermal gradient from above-average values (1·4°F/100 ft) on the basin margin, to less than average in the present Mississippi delta depocentre. Considered in detail, the picture is more complex than this, and Jam et al.[12] have shown that major gas fields are aligned along 'hot spots' which have above-average geothermal gradients. These hot spots are related to earlier depocentres of the Mississippi delta. Dailly[4] has inferred that this correlation is due to the sealing effect of moat clays beneath each depocentre. The geothermal data suggest, however, that the clays are significant, not only because they seal in hydrocarbons, but also because they seal in heat. Abnormally high pressures and temperatures are present in the deeper part of the delta,[13] and it may be that the over-pressured shales, whose thermal conductivity is low, act as insulating blankets.

Nwachukwu[21] has shown that the Niger delta, like the Gulf Coast, has sub-normal geothermal gradients in the depocentre and supra-normal gradients around the delta margin. As we have seen already, Dailly[4] has suggested a correlation between the prolific producing zones of the delta and earlier depocentres, with the 'moat' shales acting as regional seals to hydrocarbon migration. As with the Mississippi delta, it is possible that these shales are significant heat seals, but there are no published data to check this.

This review is brief, but is sufficient to draw attention to the interplay of geothermal gradient, facies and hydrocarbon generation.

CONCLUSION

The deltaic process is a means for placing porous reservoir sands far out in marine sedimentary basins. This process contains a built-in mechanism for depositing lobes of sediment with upward-coarsening grain-size profiles in which potential source beds underlie potential reservoirs.

The geometry and distribution of reservoir sands are related to the relative influence of fluvial, wave and tidal processes operating on the delta topset.

Deltaic deposits may be identified and correlated in the subsurface from seismic data and from wireline logs, as well as from sedimentary core studies.

Deltaic facies are major hydrocarbon provinces, because all five of the necessary conditions for hydrocarbon generation and entrapment are fulfilled.

There are abundant source beds whose humic nature tends to make deltas rich in gas rather than in oil. Oils are commonly heavy, but low in sulphur.

Deltas contain a diversity of reservoir sands ranging from fluvial and deltaic channel, through delta-front marine sands, to deep-sea channel and fan sands at the foot of some deltas.

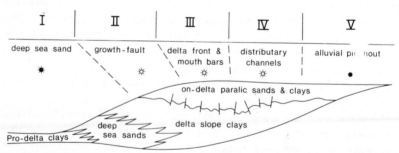

FIG. 9. Cross-section through a hypothetical delta showing hydrocarbon fairways. Up-dip pinch-out of fluvial sands may be oil-bearing due to sourcing from transgressive marine shales. Intra-deltaic traps may contain gas from humic source beds. Deep-sea play may be oil-prone owing to marine source beds, but the initial deep situation may be coupled ultimately with deep burial and high temperatures which favour gas generation.

There are adequate seals to prevent hydrocarbons migrating up-dip and being dissipated at the surface.

Deltas contain a multiplicity of hydrocarbon traps ranging from structural traps, generated by basement tectonics, to diapiric, roll-over and stratigraphic traps, which are formed by the delta itself.

Geothermal gradients in deltas range from average in the depocentre to above average on the periphery. The insulating effect of low-conductivity over-pressured shales allows the build-up of abnormally high pressures, and temperatures which are conducive to the generation of hydrocarbons in large quantities (Fig. 9).

Thus, it can be well seen why deltas contain large reserves of oil and gas, and that the research effort put into understanding these deposits has not been misplaced.

REFERENCES

1. CAMPBELL, R. L. (1968). Stratigraphic applications of dipmeter data in Mid-Continent, *Am. Assoc. Petrol. Geologists Bull.*, **52**, 1700–19.
2. COLEMAN, J. M. and GAGLIANO, S. M. (1965). Sedimentary structure: Mississippi River deltaic plain, in: *Primary Sedimentary Structures and their Hydrodynamic Interpretation*, Soc. Econ. Pal. Min, Spec. Publ., 12, 133–48.
3. COLEMAN, J. M., GAGLIANO, S. M. and SMITH, W. C. (1970). Sedimentation in a Malaysian high tide tropical delta, in: *Deltaic Sedimentation Modern and Ancient* (J. P. Morgan and R. H. Shaver, eds.), Soc. Econ. Pal. Min. Spec. Publ. 15, 185–97.
4. DAILLY, G. C. (1976). Pendulum effect and Niger Delta prolific belt, *Am. Assoc. Petrol. Geologists Bull.*, **60**, 1543–9.
5. EAMES, T. D. (1975). Coal rank and gas source relationships—Rotliegendes reservoirs, in: *Petroleum and the Continental Shelf of North-west Europe* (A. W. Woodland, ed.), Applied Science Publishers, Barking, 191–204.
6. FISHER, W. L., BROWN, L. F., SCOTT, A. J. and McGOWEN, J. H. (1972). *Delta Systems in the Exploration for Oil and Gas*, Bureau Econ. Geol. Texas Univ.
7. FISK, H. N. (1955). Sand facies of Recent Mississippi delta deposits, *Proc. 4th World Petroleum Congress*, Rome. Section 1, 377–98.
8. GILREATH, J. A. and STEPHENS, R. W. (1971). Distributary front deposits interpreted from dipmeter patterns, *Trans. Gulf Coast Assoc. Geol. Soc.*, **21**, 233–43.
9. GOULD, H. R. (1970). The Mississippi delta complex, in: *Deltaic Sedimentation Modern and Ancient* (J. P. Morgan and R. H. Shaver, eds.), Soc. Econ. Pal. Min., Spec. Publ., 15, 3–30.
10. GREGORY, J. L. (1966). A Lower Oligocene delta in the subsurface of south-eastern Texas, *Trans. Gulf Coast Assoc. Geol. Soc.*, **16**, 227–41.
11. HARMS, J. C. and TACKENBERG, P. (1972). Seismic signatures of sedimentary models, *Geophysics* **37**, 45–58.

12. JAM, L., DICKEY, P. A. and TRYGGVASON, E. (1969). Subsurface temperature in south Louisiana, *Am. Assoc. Petrol. Geologists Bull.*, **53**, 2141–9.
13. JONES, P. H. (1969). Hydrodynamics of geopressure in the northern Gulf of Mexico basin, *J. Petrol. Tech.*, **21**, 803–10.
14. KLEMME, H. D. (1975). Geothermal gradients, heatflow and hydrocarbon recovery, in: *Petroleum and Global Tectonics* (A. G. Fisher and S. Judson, eds.), Princeton Univ. Press, 251–304.
15. KOLB, C. R. and VAN LOPIK, J. R. (1966). Depositional environment of the Mississippi River deltaic plain—south-eastern Louisiana, in: *Deltas in their Geologic Framework* (M. L. Shirley and J. A. Ragsdale, eds.), Houston Geol. Soc., 17–62.
16. LAPLANTE, R. E. (1974). Petroleum generation in Gulf Coast Tertiary sediments, *Am. Assoc. Petroleum Geologists Bull.*, **58**, 1281–9.
17. LYONS, P. L. and DOBRIN, M. B. (1972). Seismic exploration for stratigraphic traps, in *Stratigraphic Oil and Gas Fields* (R. E. King, ed.), Am. Assoc. Petrol. Geologists Mem., 16, 225–43.
18. MCDANIEL, G. A. (1968). Application of sedimentary directional features and scalar properties to hydrocarbon exploration. *Am. Assoc. Petrol. Geologists Bull.*, **52**, 1689–99.
19. MORGAN, J. P. (1970). Depositional processes and products in the deltaic environment, in: *Deltaic Sedimentation Modern and Ancient* (J. P. Morgan and R. H. Shaver, eds.), Soc. Econ. Pal. Min., Spec. Publ. 15, 31–47.
20. MORGAN, J. P. and SHAVER, R. H. (eds.) (1970). *Deltaic Sedimentation Modern and Ancient*, Soc. Econ. Pal. Min., Spec. Publ. 15, 312 pp.
21. NWACHUKWU, S. O. (1976). Approximate geothermal gradients in Niger Delta sedimentary basin, *Am. Assoc. Petrol. Geologists Bull.*, **60**, 1073–7.
22. PARKER, J. R. (1975). Lower Tertiary sand development in the Central North Sea, in: *Petroleum and the Continental Shelf of North-west Europe* (A. W. Woodland, ed.), Applied Science Publishers, London, 447–54.
23. PATIJN, R. J. H. (1964). Die Entstenhung von Erd gas, *Erdol Kohle Erdgas Petrochem.*, **17**, 2–9.
24. PUSEY, W. C. (1973). Paleo-temperatures in the Gulf Coast using the ESR-kerogen method, *Trans. Gulf Coast Assoc. Geol. Soc.*, **23**, 195–202.
25. SELLEY, R. C. (1974). Environmental analysis of subsurface sediments, *Trans. Third European Symp. SPWLA*, 16 pp.
26. SELLEY, R. C. (1975). Subsurface diagnosis of deltaic deposits with reference to the northern North Sea, in: *Proc. Jurassic North Sea Conference*, Norwegian Petroleum Soc. 21 pp.
27. SELLEY, R. C. (1976a). Environmental analysis of subsurface sediments. Reprinted in *The Log Analyst*, **17**, 3–12.
28. SELLEY, R. C. (1976b). Subsurface analysis of North Sea sediments, *Am. Assoc. Petrol. Geologists Bull.*, **60**, 184–95.
29. SHIRLEY, M. L. and RAGSDALE, J. A. (eds.) (1966). *Deltas in their Geological Framework*, Houston Geol. Soc., 251 pp.
30. SHORT, K. C. and STAUBLE, A. J. (1967). Outline of geology of Niger Delta, *Am. Assoc. Petrol. Geologists Bull.*, **51**, 761–79.

Chapter 7

DEEP-SEA SANDS

J. R. PARKER

Shell U.K. Exploration and Production, London, UK

SUMMARY

This chapter considers possible models for deep-water sand deposition and relates them to examples from both past and present. Modern deep-water fans are almost invariably associated with canyons, and a depositional model can be developed in which sand is supplied through channels and canyons on an essentially non-depositional slope to the basin floor, where it accumulates as a submarine fan. Ancient examples of this canyon type of fan are found principally in tectonic settings. A second model for deposition of sand in deep water, occurring principally in non-tectonic settings and poorly represented by modern examples, is one in which sand is derived from the upper, predominantly sandy, unstable part of a prograding deltaic slope and moves down the slope by a combination of slumping, sliding, mass flow and turbidity currents. Deposition of sand occurs further down the slope to form a slope wedge in which the amount of sand can be considerable. Characteristics of these two models are discussed together with criteria for the interpretation of a deep-water environment of deposition.

INTRODUCTION

The concept of deposition of sediment in deep water by turbidity currents is one of the most significant to have influenced geological thought this century. At an early stage this concept was applied to the interpretation of the Pliocene oil reservoir rocks of the Ventura Basin in California, and since then deep-marine sandstone reservoirs have been recognised

225

elsewhere in the world. Although at present containing only a small proportion of the world's oil reserves, exploration for such reservoirs is likely to increase as exploration moves into deeper water settings.

For the petroleum geologist involved in the exploration of an offshore basin, one of the main concerns is the interpretation of the mode of deposition of a particular sedimentary unit; this interpretation frequently has to be made at an early stage in the exploration history of a basin when there is only very limited information available, such as the geometry of the sedimentary units known from seismic surveys, together with sparse well control. Based on this initial interpretation, a conceptual model is then applied to direct what is always a far too limited further exploration effort in determining the distribution of the reservoir. As more information becomes available the model is updated or perhaps abandoned.

For continental and shallow marine environments, such as deltas or barriers, where both the processes and deposits are directly observable, the models developed from recent sediments have to a large extent been successfully applied to the interpretation of ancient deposits. For deep-water environments the problem is less straightforward. The processes themselves are not directly observable and have to be inferred from circumstantial evidence and from experiments; also recent deep-sea deposits and their settings are not readily relatable to the considerable diversity of ancient deposits interpreted as having been laid down in deep water from turbidity currents. This discrepancy between past and present can be partially attributed to the lack of direct observation, but also it seems that in the case of deep-water deposition the present is only a partial replica of the past.

This review considers possible models for deep-water sand deposition and relates them to examples from both past and present.

The historical development of concepts about turbidites and deep-water deposition is traced in a review by Walker[1] and a collection of papers edited and commented on by Whitaker;[2] the latter has an extensive bibliography of examples of both modern and ancient deep-water fans.

PRESENT DAY EXAMPLES OF DEEP-WATER FANS

Studies of modern deep-sea fans are mainly concerned with the morphology of the fan and the surface sediment distribution. Numerous examples have been described, particularly from offshore California (La Jolla fan;[3,4] San Lucas fan;[4] Redondo fan[5]), and based on these

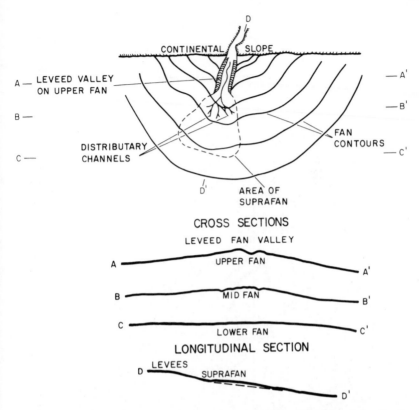

FIG. 1. Model for deep-sea fan growth showing supra-fan and simple morphology of fan related to growth area. Upper fan is characterised by large leveed fan-valley (A–A') whereas mid fan (or supra-fan) shows many smaller distributary channels and some channel remnants or depressions (B–B'). Channels are absent from lower fan (C–C'). Supra-fan appears as convex-upward bulge on a longitudinal section. From Normark.[4]

California examples a model for the growth of modern deep-water fans has been developed by Normark[4,6] (Fig. 1). The main features on the surface of modern fans are distributary channels with levees, inter-levee areas and depositional lobes below the mouths of the channels.

On the *inner* (or upper) fan, bulk sand transportation and deposition are restricted to elongate channels, often with persistent levees which may be recognised on seismic cross-sections through ancient fans. A channel on the inner fan may be entrenched below the general level of the fan surface or it may be aggrading with its floor built up above the normal fan surface. The

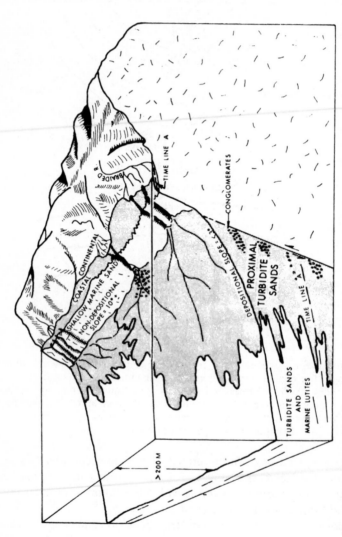

FIG. 2. Schematic illustration of facies distribution in the case of slope instability resulting from diastrophic processes. In this example, the proximal turbidites are separated from shallow marine sands by a non-depositional slope. From van der Kamp *et al.*,[7] reproduced by permission of the Geological Society of London.

relief of such a channel and the height of its levees generally decrease down the fan.

On the *middle* part of the fan numerous meandering and upbuilding channels occur. These channels lack persistent levees and pass down-current into a braided and rapidly migrating system that disappears down-fan. Rapid deposition within these channels, and just below their ends, accelerates up-building on the middle part of the fan, forming depositional lobes ('supra-fans') on the fan surface. These supra-fans appear as low, convex-upward segments on an otherwise concave-upward profile and again may be recognised on seismic cross-sections through ancient fans. Overspill deposition may occur in the inter-levee areas of both the inner and middle fans.

The *outer* (or lower) part of the fan is characterised by broad sheet-like flows and the absence of channels. It grades into the basin plain where slowly accumulating pelagic sediments alternate with thin-bedded, fine-grained, laterally-persistent turbidity flow deposits.

Since high deposition rates on the fan are restricted to channels and depositional lobes, large fans are the results of migrating sites of deposition.

Present-day deep-water fans are almost invariably associated with canyons, and a depositional model (Fig. 2) can be developed in which sand is supplied, through channels and canyons on an essentially non-depositional slope, to the basin floor, where it accumulates as a submarine fan.[7] The slopes are often tectonically controlled. Sediment is derived from deltaic and other clastic coastal complexes, and is often moved to the heads of canyons by longshore currents. Some later infilling of the channels by sand may occur, and the inter-canyon slopes may be the site of deposition of fine-grained sediment.

Submarine canyons are a very prominent feature of the present-day continental margin and appear to be related to major fluctuations in sea level during the Pleistocene. Thus, fans of the present-day California type would be expected to be developed during periods of sea level fluctuation, either of climatic or tectonic origin.

EXAMPLES OF ANCIENT DEEP-WATER FANS

Examples of the canyon type of fan discussed above have been described from the Tertiary of California (for example references 7–10; the paper by Nelson and Nilson[10] gives a comparison between a modern and an ancient deep-sea fan from offshore and onshore California) and from the Mediterranean area;[11-13] both are areas in which sedimentation has been influenced by tectonics.

Based principally on examples from the Apennines, a sediment distribution model for both ancient and modern fans of this type, together with criteria that can be used for the recognition and interpretation of the main turbidite facies and facies associations, has been developed by Mutti and Ricci Lucchi,[11.14] Walker and Mutti[15] and Ricci Lucchi.[16] The sequences shown in Fig. 3 are those which would be encountered by a series of vertical wells drilled through an idealised prograding fan. Most of the sediments of a deep-water fan appear to be deposited from turbidity currents. Slides, slumps and mass flows are generally restricted to the relatively steep parts of the fan surface and contribute to sediment distribution mainly on the inner fan. The thickest sands accumulate on the floors of depositional channels on the inner and middle fan (where they are thick-bedded, coarse to medium grained, graded or ungraded) and in depositional lobes on the outer part of the middle fan (where the sand is usually graded and medium grained). Deposition of argillaceous sediment usually predominates in the inter-channel areas of the inner and middle fan and on the outer fan; however, in sand-rich systems, considerable amounts of sand can accumulate in these areas too.

Mutti and Ricci Lucchi[11] have recognised in vertical sections thinning/fining upward sequences and thickening/coarsening upward sequences of sand bodies. The thinning/fining upward sequences are interpreted as channel fills on the inner and middle parts of the fan, and the thickening/coarsening upward sequences as sediment lobes accumulating on the middle fan beyond the downstream end of fan channels.

A further example of this type of fan has been described by Kruit et al. [17,18] from the Eocene of northern Spain. Thick, massive sand deposits which characterise Monte Jaizkibel, east of San Sebastian, represent the topographical expression of a major 'fossil' deep-water fan of early Eocene age. The fan is approximately 15 km wide and reaches a maximum thickness of about 600 m in its central area. It was produced by a lateral influx of sand into an east–west trending basin from a focal point to the north. Leeward of the fan, the sand-transporting current directions diverted into the axial basin direction. Inter-bedded mudstones contain a deep marine (bathyal to abyssal) fauna, and many sections show thickening/coarsening upward cycles, characteristic of the outer part of the middle fan.

However, there are a large number of examples of ancient deep-water deposition (principally occurring in non-tectonic settings) to which the canyon model is not applicable, and for these a second model can be developed. In this model (Fig. 4), termed the slope-wedge model, sand is derived from the upper, predominantly sandy, unstable part of a prograding deltaic slope and moves down the slope by a combination of slumping, sliding, mass flow and turbidity currents.[7] For this to occur the slope must be steep enough for instability to develop, of the order of 3° or more. Deposition of sand occurs further down the slope to form a slope wedge in which the amount of sand can be appreciable. In this situation, no submarine fan will be formed. However, during pauses in progradation, channels may be formed on the slope and, through these, sand can be transported to be deposited as a submarine fan, as described in the canyon model above. Later the deltaic complex may again prograde to overlie the fan (Fig. 5).

This model is distinguished from a simple sandy delta front/pro-deltaic slope facies, which would be a genetic part of a delta system, by the integration of the sediment dispersal system into a channelised network with fan-like morphology, and a general segregation of slope wedges from equivalent clastic units on the shelf.[19] However, if successively deposited sand bodies overlap each other, there may be a continuous transition from deep-water to shallow-marine (coastal) sands. Present-day examples of this type of model appear to be rare; the fore-delta slope of the Magdalena is a possible example.

A depositional system of the slope-wedge type has been described by Galloway and Brown[19] from the Pennsylvanian of north central Texas. Clastic sediment was passed across a shelf through one or more prograding lobes of a fluvial deltaic system, into channels that locally breached the shelf-edge carbonate bank system; sediment was then transported on to and down a prograding slope wedge and into the basin.

Another Carboniferous example is from the Namurian of the English Pennines where Walker[20] has described a regressive sequence formed by the southwards advance of a deltaic complex into an intra-cratonic basin. This sequence shows a gradual transition from deep basinal turbidites, through slope deposits, to nearshore and fluviatile sediments. A deltaic source of the turbidites is apparent from palaeo-current directions and from the presence of many turbidite-filled channels in the slope sediments. Similar sequences are seen in the Carboniferous of western Ireland and North Devon.

The two models described above represent, to some extent, the end members of a gradational series, and the North Sea Palaeocene shows a

Fig. 3. Model of deep-water slope and fan deposition. From Kruit et al.[17]

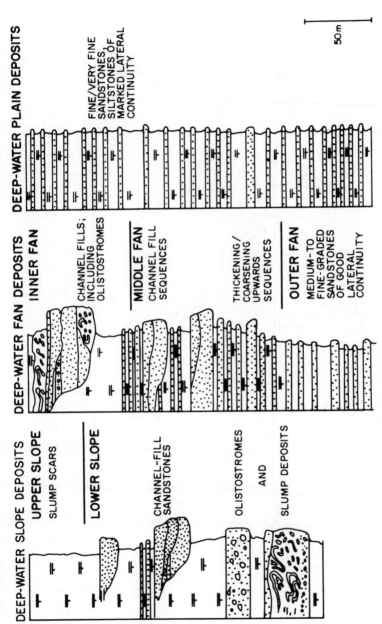

FIG. 3—contd.

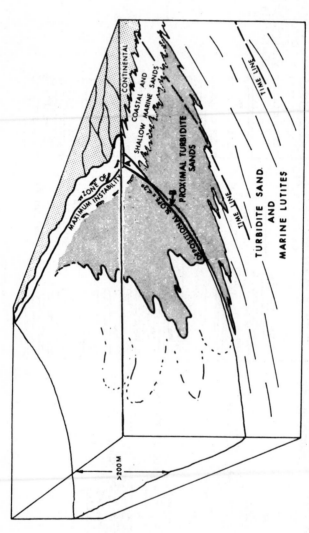

FIG. 4. Schematic representation of facies development and distribution in the case of slope instability as a function of delta building into progressively deeper water. Shallow marine sands (A) slump into deeper marine environments (B). This can result in the deposition of sand in environments ranging from bathyal to continental. From van der Kamp et al.,[7] reproduced by permission of the Geological Society of London.

transition from the major fans of the canyon model in the Lower
Palaeocene, to a slope-wedge situation in the Upper Palaeocene.[21] In early
Palaeocene times, following movements at the end of the Cretaceous, the
Outer Moray Firth and northern North Sea basins were deep-water areas
with the coast-line probably close to the present day Scottish coast. Several
fan complexes were deposited on the basin floor, and the mapped sand

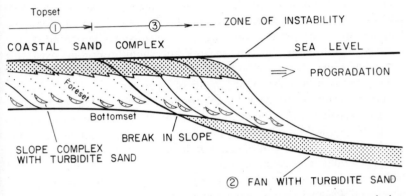

FIG. 5.   Idealised cross-section through prograding deltaic slope. (1) Progradation
of coastal sand complex to break in slope. (2) Pause in progradation; development
of fan at base of stationary slope with channelling on the slope. (3) Resumption of
progradation over fan.

distribution suggests that they were derived from separate point sources,
perhaps canyons on the basin slope. Later during the Palaeocene, a major
coastal-deltaic complex prograded eastwards over the Moray Firth area.
Sand was introduced into the basin from multiple sources along this
prograding shelf edge, with many small fans coalescing to form sand sheets
at the base of the slope.

DISCUSSION

Two models for deposition of sand in deep water have been developed: a
canyon model, primarily in tectonic settings and with variability influenced
by extra-basinal controls; and a slope-wedge model, primarily in non-
tectonic settings, and with variability resulting from the interplay of intra-
basinal depositional processes. These two models can be regarded as
representing end members of a gradational sequence, as is shown by the

North Sea Palaeocene. Although they are a considerable simplification and by no means exhaustive, the use of such models can assist the assessment of the hydrocarbon potential and reservoir distribution of a basin. In the canyon model the deep-water sand accumulation is elongated perpendicular to the slope and is separated from any shallow-water sands by the largely non-depositional slope; hence there is scope for the trapping of hydrocarbons against the base of the slope. In the slope-wedge model there is major deposition on the slope, and close to the base of the slope, and the sand accumulation is parallel to the slope. Any stratigraphic trapping of the deep-water sands against the slope would depend on the absence of continuous sand up the slope but, should continuity exist, this can act as a migration path on to the shelf for hydrocarbons generated in the organic-rich fine-grained sediments of the basin.

Interpretation of a deep-water environment of deposition can be made using core, palaeontological and log data from wells. In cores, sedimentary structures typical of shallow-water deposition (e.g. oscillation ripple marks) are virtually absent, whereas grading, load casts, contorted bedding and groove marks are common in the thinner beds (Fig. 6). Although graded bedding is frequent, the classic Bouma sequence is rarely, if ever, fully developed. Dish structures, which seem to be the result of upward moving fluidised sediment-water slurries in under-consolidated laminated sands,[22] can be a characteristic feature of the thicker beds. Liquefaction, resulting in a homogeneous sandstone, may occur following deposition, but homogeneity can also be a depositional feature. Mudstones within and around a fan may contain deep-water faunas, and characteristic trace fossils are found both on the bedding planes of sandstones and within intercalated mudstones; in contrast, the sandstones may contain derived shallow-water faunas. Idealised gamma-ray log shapes for the various parts of the deep-water fan are shown in Fig. 7. The inter-bedded thin sands and shales of the outer fan have a characteristic serrated appearance. The depositional lobes of the middle fan have thickening/coarsening upward profiles with sharp tops; the channel deposits of the middle and inner fans and the slope have thinning/fining upward profiles with sharp bases. Logs of deep-water sands of the North Sea Forties field are illustrated by Walmsley,[23] and Selley[24] presents an interpretation of a North Sea well, integrating both core and log data.

None of these features in isolation are specifically characteristic of deep-water sands, and interpretation should be based on as much evidence as possible. The interpretation of deep-water environments solely on the basis of log characteristics should be made with great caution, since similar

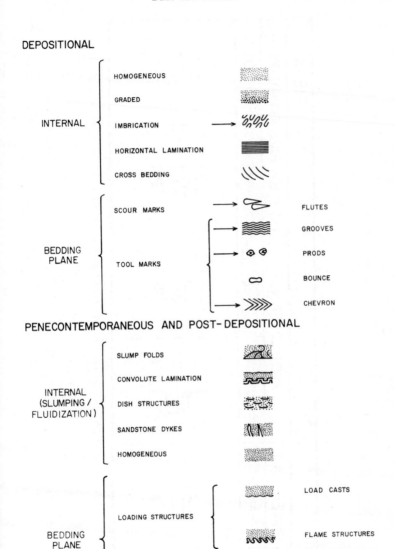

FIG. 6. Common sedimentary features of deep-sea sands. Arrow indicates current direction.

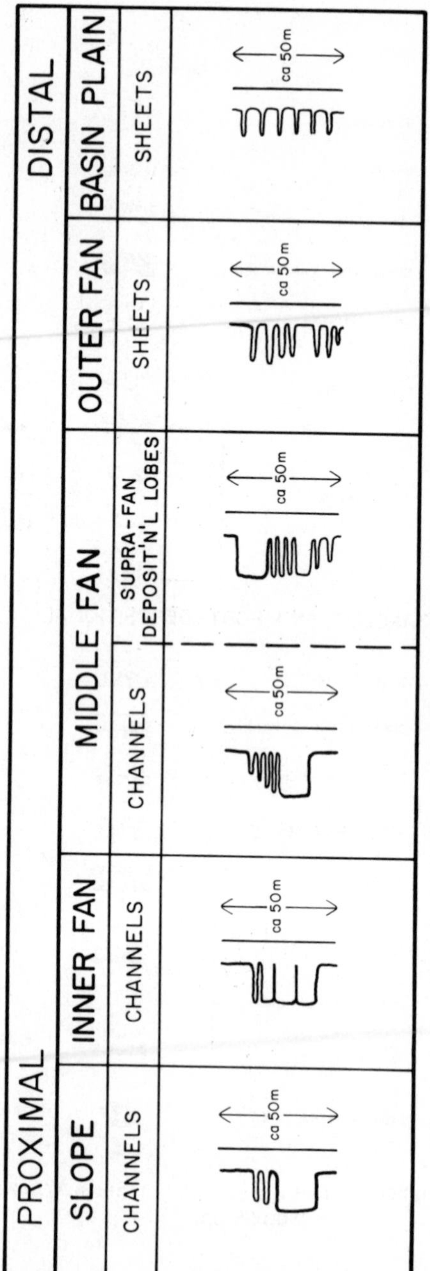

J. R. PARKER

Fig. 7. Idealised gamma-ray log shapes of turbidite sequences.

SUMMARY OF CHARACTERISTICS OF DEEP-SEA SANDS

	PROXIMAL				DISTAL	
	Slope	Inner fan	Middle fan		Outer fan	Basin Plain
Main features	slope wedge slope channels and canyons	channels levees inter-levee areas	channels braided distributary system levees inter-levee areas	depositional lobes	sheets	sheets
Depositional processes	slides slumps mass flows turbidity currents (limited)	decreasing amount of slumped sediment in fan channels predominantly turbidity currents pelagic deposition in inter-levee areas			turbidity currents pelagic deposition	
Bed thickness	thick bedded <1 m–10 m, thickest in channel axis; amalgamated sand beds to 100 m	thick bedded channel fills	local deep channels thinning/fining upward cycles (20–60 m)	local shallow channels thickening/coarsening upward cycles (40–200 m)	thin-bedded (<0.5 m)	
Bed continuity	slope wedge; laterally continuous over several km. slope channels and canyons: continuous down channel	continuity better in direction of depositional dip				continuous over large area
Grading	ungraded–graded Bouma ae	Bouma ae (ab)	graded most complete Bouma abcde		Bouma bcde/cde	Bouma cde/de
Grain size and sorting			grain-size fining → sorting improving →			

shapes are found in shallow-water sequences. Identification of sedimentary structures in cores and outcrops is probably the most reliable single method for the recognition of ancient deep-water deposits. For the interpretation of a depositional model it is then necessary to relate the well data to the regional tectonic setting and to the mapped distribution of the sedimentary units: for example, the thick accumulation of sediment on the slope in the slope-wedge model is often a very striking feature on seismic cross-sections.

The depths at which ancient deep-water fans were deposited are difficult to define precisely. The absence of shallow-water sedimentary structures sets a minimum depth below wave base of. say. at least 50 m. One approach is to consider the reconstructed geometry of shelf, slope and basin (undaform, clinoform and fondoform of Rich[25]). For the Pennsylvanian of north central Texas,[19] the preserved relief between the shelf margin and the basin floor ranges from 600 to 1100 ft (180–330 m), with dips ranging up to 5°. For the North Sea Palaeocene, the relief is of the order of 2500 ft (760 m); here foresetting on a huge scale, representing the slope, can be recognised on seismic sections.[21] Alternatively. and less precisely. faunal evidence from intercalated shales can be used: this suggests water of bathyal depth, that is, probably in the order of 600–3000 ft (180–900 m) for the North Sea Palaeocene[21] and somewhat deeper (at least 1000 m) for the San Sebastian Eocene.[17.18] Table 1 gives a summary of the characteristics of deep-sea sands.

ACKNOWLEDGEMENT

Permission for publication was given by Shell UK Exploration and Production, and is gratefully acknowledged.

REFERENCES

1. WALKER, R. G. (1973). Mopping up the turbidite mess, in: *Evolving Concepts in Sedimentology* (R. N. Ginsburg, ed.), The Johns Hopkins University Press, Baltimore and London.
2. WHITAKER, J. H. MCD. (1976). *Submarine Canyons and Deep-Sea Fans, Modern and Ancient*, Dowden, Hutchinson and Ross, Inc., Stoudsburg.
3. SHEPARD, F. P., DILL, R. F. and VON RAD, U. (1969). Physiography and sedimentary processes of La Jolla submarine fan and fan-valley, California, *Am. Assoc. Petrol. Geologists Bull.*, **53**, 390–429.

4. NORMARK, W. R. (1970). Growth patterns of deep-sea fans, *Am. Assoc. Petrol. Geologists Bull.*, **54**, 2170–95.
5. HANER, B. E. (1971). Morphology and sediments of Redondo submarine fan, southern California, *Bull. Geol. Soc. Am.*, **82**, 2413–32.
6. NORMARK, W. R. (1974). Submarine canyons and fan valleys: factors affecting growth patterns of deep-sea fans, in: *Modern and Ancient Geosynclinal Sedimentation* (R. H. Dott and R. H. Shaver, eds.), Soc. Econ. Pal. Min., Spec. Publ. 19.
7. VAN DER KAMP, P. C., HARPER, J. D., CONNIFF, J. J. and MORRIS, D. A. (1974). Facies relations in the Eocene–Oligocene in the Santa Ynez Mountains, California, *J. Geol. Soc. Lond.* **130**, 545–65.
8. SHELTON, J. W. (1967). Stratigraphic models and general criteria for recognition of alluvial, barrier-bar and turbidity-current sand deposits, *Am. Assoc. Petrol. Geologists Bull.*, **51**, 2441–61.
9. PIPER, D. J. W. and NORMARK, W. R. (1971). Re-examination of a Miocene deep-sea fan and fan-valley, southern California, *Bull. Geol. Soc. Am.*, **82**, 1828–30.
10. NELSON, C. H. and NILSON, T. H. (1974). Depositional trends of modern and ancient deep-sea fans, in: *Modern and Ancient Geosynclinal Sedimentation* (R. H. Dott and R. H. Shaver, eds.), Soc. Econ. Pal. Min., Spec. Publ. 19.
11. MUTTI, E. and RICCI LUCCHI, F. (1972). Le torbiditi dell'Apennino settentriole: introduzione all' analisi de facies, *Mem. Soc. Geol. Italiana*, **11**, 161–99.
12. MUTTI, E. (1974). Examples of ancient deep-sea fan deposits from circum-Mediterranean geosynclines, in: *Modern and Ancient Geosynclinal Sedimentation* (R. H. Dott and R. H. Shaver, eds.), Soc. Econ. Pal. Min., Spec. Publ. 19.
13. RICCI LUCCHI, F. (1975). Miocene paleogeography and basin analysis in the Periadriatic Apennines, in: *Geology of Italy* (C. Squyres, ed.), Petroleum Exploration Society of Libya, Tripoli.
14. MUTTI, E. and RICCI LUCCHI, F. (1975). Turbidite facies and facies associations, in: 'Examples of Turbidite Facies and Facies Associations from Selected Formations of the Northern Apennines', *Field Trip A 11, IX International Congress of Sedimentology*, Nice.
15. WALKER, R. G. and MUTTI, E. (1973). Turbidite facies and facies associations, in: *Turbidites and Deep Water Sedimentation*, Soc. Econ. Pal. Min., Short Course notes, 119–57.
16. RICCI LUCCHI, F. (1975). Depositional cycles in two turbidite formations of northern Apennines (Italy), *J. Sediment. Petrol.*, **45**, 3–43.
17. KRUIT, C., BROUWER, J., KNOX, G., SCHOLLNBERGER, W. and VAN VLIET, A. (1975). Une excursion aux cônes d'alluvions en eau profonde d'âge Tertiare près de San Sebastian, *Guide pour l'excursion Z23, IX Congrès International de Sedimentologie*, Nice.
18. KRUIT, C., BROUWER, J. and EALEY, P. (1972). A deep-water sand fan in the Eocene Bay of Biscay, *Nature Phys. Sci.*, **240**. 59–61.
19. GALLOWAY, W. E. and BROWN, L. F. (1973). Depositional systems and shelf-slope relations on the cratonic basin margin, Upper Pennsylvanian of north central Texas, *Am. Assoc. Petrol. Geologists Bull.*, **57**, 1185–1218.

20. WALKER, R. G. (1966). Shale Grit and Grindslow Shales: transition from turbidite to shallow water sediments in the Upper Carboniferous of northern England, *J. Sediment. Petrol.*, **36**, 90–114.

21. PARKER, J. R. (1975). Lower Tertiary sand development in the central North Sea, in: *Petroleum and the Continental Shelf of North-west Europe*, **1** (A. W. Woodland, ed.), Applied Science Publishers, Barking.

22. LOWE, D. R. and LO PICCOLO, R. D. (1974). The characteristics and origins of dish and pillar structures, *J. Sediment. Petrol.*, **44**, 434–501.

23. WALMSLEY, P. J. (1975). The Forties field, in *Petroleum and the Continental Shelf of North-west Europe*, **1** (A. W. Woodland, ed.), Applied Science Publishers, Barking.

24. SELLEY, R. C. (1976). Subsurface environmental analysis of North Sea sediments, *Am. Assoc. Petrol. Geologists Bull.*, **60**, 184–95.

25. RICH, J. L. (1951). Three critical environments of deposition and criteria for recognition of rocks deposited in each of them, *Bull. Geol. Soc. Am.*, **62**, 1–19.

Chapter 8

USING SEISMIC DATA TO DEDUCE ROCK PROPERTIES

R. E. SHERIFF

Seiscom Delta Inc., Houston, Texas, USA

SUMMARY

The desire to know the nature of the rocks which may host petroleum accumulations usually poses more questions than answers. These questions often are about details such as porosity and permeability of a very small portion of the sedimentary section. We have to infer answers from the measurements we make, being aware of the uncertainties in our measurements and in interpretations from those measurements.

Two types of approach are open to us: (1) to get close to the rocks we want to study, or (2) to sense their properties from a long way off. The first approach involves drilling holes and lowering instruments in them. It has two major disadvantages: (a) a hole allows us to look at only a very small amount of rock around the borehole, and (b) the cost of drilling and logging holes is very large. The second approach allows us to look at an entire region, but with less precision. The two approaches answer different questions and so are not in competition. By combining measurements on the surface with measurements in boreholes, we can obtain more useful information than from either approach by itself.

Seismic reflection measurements are the principal type of surface measurements from which rock properties are deduced. The depth of penetration and resolving power of seismic waves generally exceed that of other types of surface measurements. Measurements of different kinds, such as the gravity or magnetic fields, or surface geology, help to answer other questions. Incorporating such with seismic measurements allows a more complete interpretation.

243

Appreciation of the reliability of rock property determinations requires an understanding of how seismic measurements are made and how the variabilities in field situations affect measurements. Hence, a discussion of seismic limitations precedes discussions of synthetic seismograms, seismic logs, attribute analysis, seismic facies analysis and velocity analysis methods of analysing seismic data with the objective of the deduction of rock properties.

TYPES OF SEISMIC MEASUREMENTS

Seismic data must be recorded and processed in such a manner that it can be assumed that the data represent primary reflections in correct amplitude relationships, spaced appropriately to the subsurface locations of the interfaces with which they are associated. The steps by which this is accomplished are vital parts of the interpretation process, and errors in these steps will prejudice subsequent interpretation.

Interpretation then takes on one of two forms: (1) picking and making measurements on individual reflections, and/or (2) analysing the sequence and pattern of successive reflections.[12] Measurements of individual reflections include arrival time, magnitude, polarity, lateral continuity, dip and normal move-out (stacking velocity). Measurements of the sequence of reflections involve their spacing, relative abundance, relative amplitude and relative attitude. The distinction between individual reflection and pattern measurements is not always clear, since successive reflections may blend together in ways which are evidenced by changes in wave shape or amplitude.

SEISMIC RESOLVING POWER LIMITATIONS

Whether features can be seen in their effect on seismic waves depends on their magnitude compared with the wavelength. Seismic waves involve a distribution of wavelengths; one usually thinks in terms of a pass-band of frequencies which contains most of the energy. Frequency bears a reciprocal relationship to wavelength:

$$\text{Wavelength} = (\text{Velocity})/(\text{Frequency})$$

The wavelengths involved in conventional seismic exploration are in the general range of 30 to 300 m and even 'high resolution' usually does not shorten these values by more than half. Wavelength generally increases with depth, because (a) velocity generally increases, and (b) higher frequencies attenuate more rapidly so that the frequencies become lower. Since wavelength limits resolving power, deep features have to be much larger than shallow features to produce similar seismic expression.

Resolution is the ability to distinguish separate features. Resolution is usually expressed as the minimum distance between two features such that one can tell that there are two features rather than one. Seismic interpretation is concerned with resolution in two directions: vertically in arrival time and horizontally from trace to trace.[14]

Vertical resolution is concerned with how far interfaces have to be separated so that it is clear that there is more than one interface. Since beds are seen by reflections from their upper and lower surfaces, vertical resolution is concerned with how thin a bed may be before it becomes indistinguishable. Horizontal resolution depends on how far apart horizontally features have to be to give separable effects, that is, how small areally a feature can be before it becomes indistinguishable.

Resolution is somewhat subjective, depending on background noise, the interpreter's sensitivity to minor wave-shape changes, etc. Generally, vertical resolution is of the order of one tenth to one quarter of the dominant wavelength, and horizontal resolution is of the order of the Fresnel zone diameter. Under ideal conditions, such as noise-free sample situations where a good reference is available, resolution can be smaller than this.

A distinction should be drawn between resolving power limitations and the accuracy of measurements. Arrival time is usually measured to the millisecond, although only rarely with better than a few milliseconds' accuracy. In some areas anomalies of only a few milliseconds' magnitude are indicative of hydrocarbon traps. Since depth determination depends on both the accuracy of timing and of velocity, uncertainty in either will cause uncertainty in depth calculation. Two to five metres of depth uncertainty results from either 2 ms of timing uncertainty or a velocity uncertainty of 4 m/s. Errors because of velocity uncertainty may exceed errors due to time uncertainty. Distinction should also be made between the ability to determine the horizontal position of a feature, such as a fault, which may be as small as one or two times the trace spacing (usually 50 m), and the distinction as to whether more than one feature is involved.

Huygens' principle implies the concept that something happens physically at a reflecting interface that generates a reflection. Seismic waves involve particles moving with respect to neighbouring particles, which changes the elastic force between them and in turn produces movement of the neighbouring particles. The net effect of the motion of nearby particles is the sum of the individual effects of each of them; the effects of some counter the effects of others. A Fresnel zone constitutes all the contiguous particles whose effects unite constructively. With specific reference to a reflecting interface, it is the area from which energy can be reflected such that the distance the energy travels from source to detector is within a half-wavelength variation, so that interference is constructive. The magnitude of a Fresnel zone can be illustrated by a few examples. For 60 Hz energy from a point source and a plane reflecting interface at a depth of 1000 m where the average velocity above the reflector is 2000 m/s, the Fresnel zone has a radius of 130 m; for 30 Hz it is 183 m. For a deeper reflector at 4000 m with an average velocity of 3500 m/s, the first Fresnel zone has a radius of 375 m for 50 Hz, and 594 m for 20 Hz.

A quarter to a tenth of a wavelength is often the limit of resolving power. Consider the gradual pinch-out of a sand wedge with the same shale above and below, as shown in Fig. 1. If the wedge is thicker than a quarter of a wavelength, one can tell that more than one interface is involved. For such a wedge, reinforcement occurs at a quarter of a wavelength. Below a tenth of a wavelength the amplitude of the reflection weakens rapidly. The decrease in amplitude as the wedge becomes very thin could be used to determine the

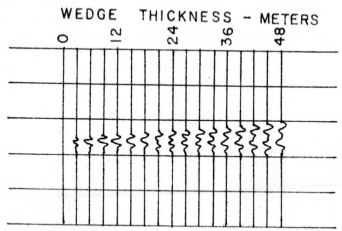

FIG. 1. Response of a wedge embedded in matter with different acoustic impedance (such as a sand pinch-out in a shale body). For the parameters assumed (50 m wavelength, i.e. 2500 m/s velocity and 50 Hz dominant frequency), the thicknesses are given in metres and the timing lines are 10 ms apart. The greatest amplitude is the 12 m trace, for which the wedge is 1/4 λ thick.

wedge thickness, if it were known that two interfaces of opposite reflection coefficients were involved and also if the amplitude for a known thickness could be determined as a reference.

Resolution can be improved by recording higher and broader bands of frequencies; this is the goal in 'high-resolution' recording. Some improvement can be obtained by using sources which are rich in high frequencies and by not discriminating against high frequencies in recording and processing. The latter involves sampling more often and avoiding the filtering which accompanies minor differences between the elements of source or detector arrays. However, often only minor improvement can be achieved at large depths. Noise also limits resolution, and the increase in noise by high resolution methods, such as by avoiding ground mix, may offset the benefit of not cancelling the high frequencies.

Spatial resolution can be effectively improved by migration, but the influence of Fresnel zone size then shows up as migration noise. Signal/noise ratio may provide the practical limitations for spatial resolution.

SEISMIC METHOD LIMITATIONS

Seismic method limitations often are only economic and can be circumvented if sufficient effort is made. Sometimes the effective depth of penetration is limiting. Depth of penetration depends on

(1) getting enough seismic energy into the Earth so that reflections stand out in comparison with background seismic energy, and
(2) having reflections from the desired portion of the Earth sufficiently strong in comparison with other coherent seismic energy.

Field and processing techniques can help in both regards. More seismic energy can be fed into the Earth by the use of stronger sources or by injecting energy over a longer period of time, as by using a long Vibroseis sweep or by stacking more records together. Arrays of sources or detectors and their layout can be used to discriminate against noise.

The attenuation of noise is usually the principal objective of seismic data processing. Noise can be attenuated by discriminating on the basis of aspects in which desired reflections differ from undesired noise modes. The key in developing methods to remove noise effects is understanding noise sources and then processing to remove the effects of specific noise sources by exploiting the characteristics of the sources. Signature-correction processing can correct for source wave-shape variations when these are known. The redundancy provided by recording with high degrees of multiplicity and velocity filtering permit the attenuation of many types of coherent wave trains, and also random noise. Predictive deconvolution and common-depth-point stacking attenuate multiples. Statistical deconvolution helps to remove near-surface reverberation and to broaden the frequency spectrum so as to sharpen the seismic wavelet.

A second important objective of processing is to position data elements in proper spatial relationships rather than underneath the locations at which the data are observed. Migration is usually thought of as a structural interpretation tool, to help clarify regions of structural complexity. Even in complex areas, migration is often not used, based on the argument that it does not alter the location of the crest of a structure and it does not matter if the picture of the flanks is distorted. Where the bedding does not depart too much from parallel, migration has generally been thought to be an unnecessary luxury. However, migration also clarifies stratigraphic evidence.

Stratigraphic features often involve elements which have slight dip with respect to the overall bedding attitude, and these must be positioned correctly for their significance to be appreciated. An example of the benefits of wave-equation migration in such a situation is shown in Fig. 2. Migration by the wave-equation method, when properly done, does not distort wave shape or amplitude, thus preserving two of the keys vital to stratigraphic interpretation.

The importance of a display so the interpreter can comprehend the many data elements on a seismic section and appreciate the significance of inter-relationships, is often underestimated.[13] Scale and scale ratio are important aspects relating to the ability to see stratigraphic features. A compressed horizontal scale is often helpful in seeing stratigraphic changes, although it may make structural interpretation more difficult, because of vertical exaggeration distortion. An enlarged display of the vicinity of critical horizons may show subtle variations that are difficult to see on a conventional section, although the latter may be better in showing the overall picture. Usually no single display is ideal for all needs and several displays should be used, each optimised for specific objectives.

Different displays are often required for seismic-derived attribute quantities such as amplitude, polarity, frequency, velocity, etc. Sometimes a section is plotted at several amplitudes, because otherwise the full range of significant amplitude changes cannot be comprehended. Likewise, a section may be replotted with reverse polarity to emphasise different

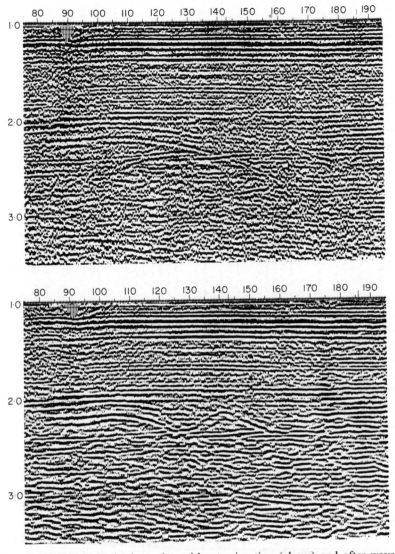

FIG. 2.　Portion of seismic section without migration (above) and after wave-equation migration (below). A salt layer is present at the left edge from about 2·2 to 2·4. This salt has been removed by solution in the vicinity of SP 125 and to the right of SP155, with a pad of residual salt in the vicinity of SP140. The salt features are much clearer after migration. Tip-off as to their presence is also indicated by variations in the interval between reflections above them. The overlying beds show the effects of the salt solution which was under way during their deposition.
(Courtesy Seiscom Delta Inc.)

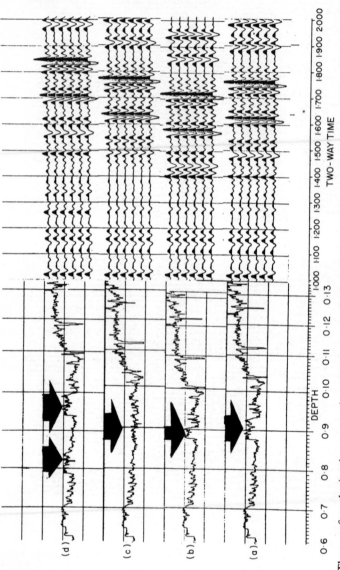

Fig. 3.   The use of synthetic seismograms in studying stratigraphic and lithological variations. The velocity log (a) is a portion of the same log shown in Fig. 4, and below it is the corresponding portion of a synthetic seismogram. The log has been modified in (b) to make the sand body thinner, and the synthetic seismogram below indicates how such a thinner sand should appear on a seismic record. In (c) the log has been modified by replacing the sand with a shale facies, and in (d) the single sand body has been replaced by two separate sands. The changes in this example are much more extreme than usually made, to illustrate the method.   Depth values are in thousand feet. (Courtesy Seismograph Service Corp.)

aspects. Colour provides an especially valuable way of conveying variations in attribute measurements to an interpreter in a way which permits greater comprehension of inter-relationships.

SYNTHETIC SEISMOGRAMS

Synthetic seismograms provide a means for linking borehole logs with actual seismic records. Their principal use is in identifying which event on a seismic record relates to a particular interface or sequence of interfaces. By comparing actual field records with synthetic seismograms made for primary reflections and for primaries-plus-multiples, one may be able to determine which events are primary reflections. Another very important use of synthetic seismograms is to see the effects of a change in the geological section. One can vary the input data to simulate changes in the thickness of units, the disappearance of units, changes in lithology, etc. (Fig. 3). This use of synthetic seismograms provides a guide to what to look for on a seismic section as evidence of changes which are expected, such as pinch-outs, channel sand development or other facies changes. Synthetic seismograms are a simple form of seismic modelling.

Changes in the product of density and velocity, called 'acoustic impedance', give rise to reflections. A synthetic seismogram involves calculating the seismic record which should be observed for a given sequence of rock units. The amplitude of a reflected wave compared to that of the incident wave is called the reflection coefficient; it depends on the properties of the material on opposite sides of an interface. If a seismic wave approaches an interface nearly broadside, the reflection coefficient is simply expressible in terms of the products of the density and velocity on opposite sides of the interface:

$$\frac{\text{Amplitude of reflected wave}}{\text{Amplitude of incident wave}} = \text{Reflection coefficient}$$

$$= \frac{\text{Change in product of density and velocity}}{\text{Twice the average product}}$$

The procedure in the generation of a synthetic seismogram (Fig. 4) is to generate a reflection coefficient log from changes in velocity and density values, and to convolve this with a seismic wavelet.

Synthetic seismograms sometimes fail to match actual seismic records and thus are not as useful as expected, because the wrong wavelet shape is

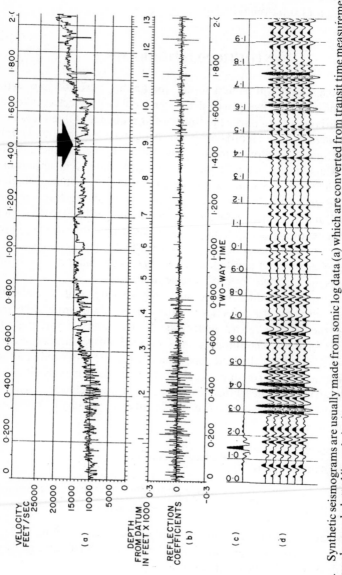

Fig. 4. Synthetic seismograms are usually made from sonic log data (a) which are converted from transit time measurements to velocity values and plotted linearly in time. Because of the variation of velocity with depth, the depth scale is more compressed for the deeper part of the borehole. A reflection coefficient log (b) is generated from changes in the velocity, and this is convolved with a wavelet (c) to create the synthetic seismogram (d). To be more correct, density data should also have been incorporated. The arrow indicates the sand body which is varied in Fig. 3. (Courtesy Seismograph Service Corp.)

assumed. Techniques for extracting wavelet shape from seismic data, for compensating for variations in source and near-surface conditions, and for transforming seismic records to what would have been expected had the wavelet been different (zero phase instead of minimum phase, for example), permit better matching by synthetic seismograms.

The most important limitation in making synthetic seismograms is that the input data usually are incomplete. Reliable density data are often missing over most of the borehole, although this may not produce severe error because density usually varies in the same way as velocity. The greatest limitation usually is that the velocity data are incomplete, especially over the upper part of the hole. Often logs are not run over the portion of a hole covered by the first casing string, a region within which major velocity variations commonly occur. This shallow portion of the Earth often includes the dominant generators of multiples, especially the base of the weathering, so that the strongest multiples may not be calculated correctly. Interference with such multiples may confuse the data in which we are particularly interested.

SEISMIC LOGS

Whereas synthetic seismograms provide a means for calculating a seismic trace from well data, a 'seismic log' is an attempt to calculate the equivalent of a sonic well log from the seismic trace. The equation given above for reflection coefficient can be solved for the acoustic impedance of the lower medium:

$$\frac{\text{Velocity–density product below interface}}{\text{Velocity–density product above interface}} = \frac{1 + \text{Reflection coefficient}}{1 - \text{Reflection coefficient}}$$

Density variations are often neglected in making seismic logs as in making synthetic seismograms. Seismic log manufacture assumes that the seismic trace contains only the complete set of primary reflections and that these are present in proper proportions. Non-reflection energy, including multiples, must have been removed, amplitude must have been preserved faithfully, and the equivalent wavelet must have been reduced as much as possible to an impulse. Since these requirements cannot be met completely, the process is at best an approximation, and its application is often very limited. However, in some instances seismic logs can be produced which are reasonable approximations to filtered velocity logs. An example of seismic logs calculated from seismic data is shown in Fig. 5.

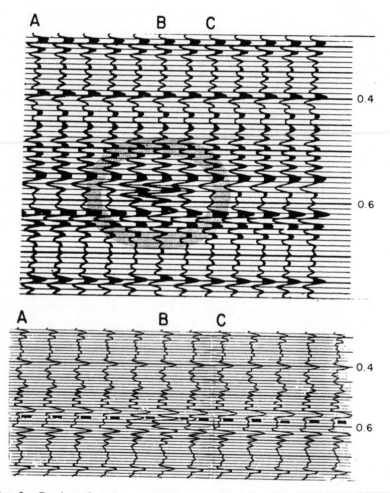

FIG. 5.   Portion of a seismic section (above) and of seismic logs made from the respective data (below). The dashes on the seismic log section locate a sand body which is 25 ft thick in a well at location A, 60 ft thick at location B, and 10 ft thick at location C. This sand produces gas in wells A and B. From Laing.[6] (Courtesy Continental Oil Company.)

SEISMIC ATTRIBUTE ANALYSIS

The concept that minor variations in the seismic wave form have significance with respect to geological variations led to the search for measurements of aspects of seismic data which are sensitive to such variations. Seismic attribute analysis involves:

(1)   making objective measurements of such aspects,
(2)   displaying these in a form that can be comprehended by an interpreter, and
(3)   attributing geological significance to them.

Velocity analysis (which is described subsequently) involves one such attribute, and amplitude measurement constitutes another. Amplitude measurements are especially used in 'bright spot' analysis for gas accumulations, which is discussed in Chapter 9. Complex trace analysis provides a useful set of measurements of seismic data attributes.[15]

The concept underlying complex trace analysis is that the seismic signal can be thought of as analytic. It can thus be decomposed into envelope amplitude and instantaneous phase terms, which are independent time variables. Standard mathematical techniques (Hilbert transform) allow us to accomplish this. The time derivative of the phase then gives the instantaneous frequency. The relationship of the phase to the envelope can also be measured; the sign of the phase angle when the envelope reaches a local maximum gives the polarity. These several measurements can also be combined in various ways; for example, the instantaneous frequency can be weighted according to the envelope and averaged to produce weighted averaged frequency. A portion of a seismic trace and attribute measurements are illustrated in Fig. 6.

A physical rationale for this type of measurement can be provided by thinking in terms of a damped harmonic system which receives additional energy at irregular times. Seismic detectors measure the velocity of motion and hence the kinetic energy. Kinetic energy involves the square of the velocity, and hence the square of the amplitude of a seismic trace. Likewise, the potential energy involves the square of the amplitude of the quadrature trace, which can be calculated from the seismic trace. The total energy in the system, the sum of the potential and kinetic energy, is thus represented by the complex trace.

Reflections on Fig. 6 arrive as fairly discrete bundles of energy with a time duration of about two cycles. The envelope (reflection strength) looks rather like beads of different sizes strung along the time axis. The phase plot

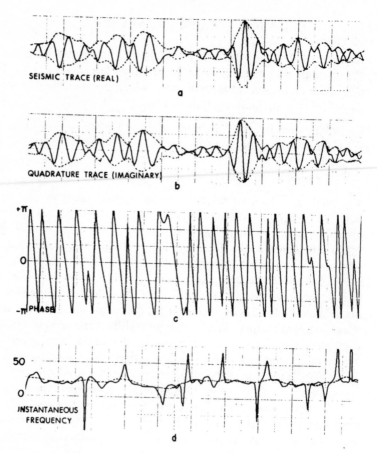

FIG. 6. Attribute analysis of a portion of a seismic trace indicated by a solid line in (a). The corresponding quadrature trace is shown in (b). The envelope ('reflection strength') is dotted in (a) and (b). The phase angle is graphed in (c), and the instantaneous frequency is indicated by the solid line in (d). Weighted averaged frequency is shown by the dotted curve in (d).

shows changes which mark the onset of each of these; these changes become spikes when differentiated to generate the instantaneous frequency.

The instantaneous phase carries the arrival time information, the mapping of which is usually the primary objective of seismic work. Thus, the attitude of an interface is given by following the same phase angle from trace to trace. A phase plot (Fig. 7) facilitates seeing the coherent aspects of

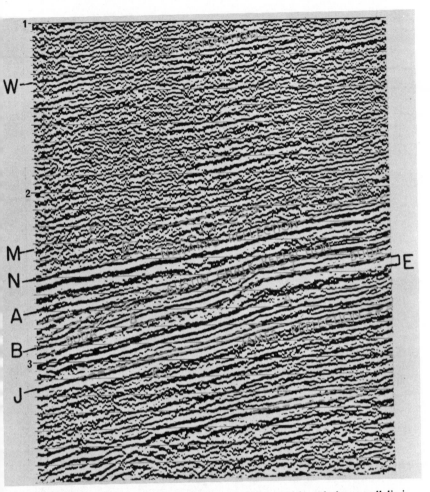

FIG. 7.   Constant phase plot of a land line in east Texas. Although the overall dip is fairly uniform over the entire section, stratigraphic features show up in the prograding Midway (between M and N) and Woodbine sections (between A and B). Other stratigraphic features include the Edwards Reef (E, extending about a third of the distance across the section), down-lap against J and elsewhere. Reflections have been identified by their usual names: W, Wilcox; M, Midway; N, Navarro; A, Austin Chalk; B, Buda limestone; J, James. The up-dip pinch-outs of the Woodbine sands produce gas in this area. Increases in the reflection strength, which do not show on the phase plot, mark the gas in these up-dip pinch-outs. (Courtesy Seiscom Delta Inc.)

the data and makes it easier to pick weak events, especially in the presence of interference from events with different attitudes. Interruptions in continuity, such as faults and pinch-outs, are especially evident on phase displays. The up-dip pinch-outs of the Woodbine sands show up in Fig. 7 with especial clarity. Features such as turbidite sand build-ups are sometimes evident on phase plots.

The reflection strength indicates the degree of contrast at interfaces. Unconformities and seismic sequence boundaries (such as will be discussed in a later section) often are marked by relatively high reflection strength. Angular unconformities are commonly characterised by variations in reflection strength along them, as the nature of the sub-cropping beds changes. The magnitude and constancy of reflection strength helps in selecting reference horizons for isopach studies, so that the number of variables to which observed interval variations can be attributed is reduced. The presence of hydrocarbons, especially gas, is sometimes shown by reflection strength.

Frequency plots, both of instantaneous and weighted frequency, provide another measure which is often useful in correlation, especially across faulting. Distinctive patterns may mark particular bedding sequences, lithologies or fluid content. Distinctive frequency patterns sometimes mark reef limestone, fractured chalk zones and hydrocarbon accumulations in certain areas.

One of the benefits in having several displays of different attribute measurements is that they show features of different kinds. Anomalies based on only one line of evidence are always suspect, especially when the data are noisy. Where several independent types of evidence each point to an anomaly, one can be reasonably certain that it exists, although the nature of the anomaly may still be subject to interpretation.

Relating anomalies in attribute measurement to situations known from well control is important in developing confidence in the meaning of measurements. Attribute measurements, when tied to well control, have the potential of showing where conditions change. They have potential especially for defining productive limits as an aid in determining how to develop a new field.

SEISMIC FACIES ANALYSIS

Seismic facies studies have become more prominent recently. The session on seismic stratigraphy at the 1975 AAPG meeting in Dallas and the

AAPG–SEG 'Stratigraphic Interpretation of Seismic Data' school in Vail, Colorado in July, 1976, gave widespread recognition to this aspect of seismic interpretation.†

'Facies' refers to 'the sum total of features, such as sedimentary rock type, mineral content, sedimentary structure, bedding characteristics, etc., which characterise a sediment as having been deposited in a given environment'.‡ Facies involves, among other things, sedimentary structure, the form of the bedding, original attitude, and the shape, thickness, variations in thickness and continuity of sedimentary units. A number of depositional patterns can be seen in seismic data (e.g. on Fig. 7), including progradation, pinch-outs and reefs. These depositional patterns sometimes can be associated with changes in depositional energy, lithology, porosity, or other physical properties which are important in hydrocarbon reservoirs.

Depositional patterns associated with significant changes in physical properties should be expected in seismic data, although many features are too small to see. Many 'stratigraphic traps' involve non-depositional features, such as angular unconformities, which may be evident in seismic data. The more obvious sedimentary patterns have been recognised for years. However, sedimentary patterns often have been obscured by superimposed noise or long wavelet length. Improvements in seismic data recording and processing have led to the present appreciation that sedimentary patterns often can be seen.

The taxonomy of seismic facies patterns was described by Sangree and Widmier.[11] Their classification of reflection configurations into parallel, divergent, oblique, sigmoid, chaotic, or reflection-free is illustrated in Fig. 8. Parallel geometry suggests uniform deposition on a stable or uniformly subsiding surface. Divergent geometry indicates areal variation in the rate of deposition, progressive tilting, or a combination of the two. The sigmoid and oblique configurations form through the progressive building out on surfaces that slope gently into deeper water. The top of the oblique pattern indicates shallow water and relatively high energy of deposition, often

---

† Papers emanating from the Exxon organisation brought the term 'seismic facies' to the forefront, such as 'Recognition of continental-slope seismic facies, offshore Texas–Louisiana' by Sangree et al.[10] Papers by others have taken up the terminology, such as 'Seismic facies reflection patterns: examples from Brazilian rift and pull-apart basins' by Brown and Fisher.[1] Both of these were part of the AAPG–SEG Stratigraphic Interpretation of Seismic Data school.

‡ As defined by D. G. A. Whitten and J. R. B. Brooks in *A Dictionary of Geology*, Penguin Books, 1972.

delta-plain sediments. The sigmoid pattern indicates continuing subsidence and low depositional energy, sometimes deposition in fairly deep water. The sigmoid pattern is often sand-poor. The chaotic geometry of discordant, discontinuous reflections suggests either relatively high energy, variability, or disruption after deposition. Sometimes this pattern, as caused by deposition or by subsequent deformation, can be distinguished

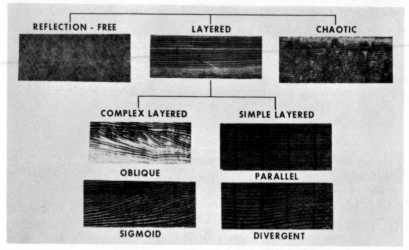

FIG. 8. Seismic facies patterns. From Sangree and Widmier.[11]

by evidences of folding, small faults, etc. A reflection-free geometry suggests a uniform lithology, such as a homogeneous marine shale in a clastic environment, or salt, or massive carbonates in an evaporite–carbonate environment.

Seismic facies with high continuity and high amplitude suggest continuous strata deposited in a relatively widespread, uniform environment. They normally indicate a neritic marine environment with shale interbedded with relatively thick sands, silts, or carbonates, although fluvial sediments with interbedded clays and coals generate the same pattern. Low-amplitude zones indicate either beds too thin to be resolved or one predominant lithological type. Low-amplitude, sand-prone seismic facies may be distinguished from low-amplitude, shale-prone facies by the landward and seaward equivalents, if these can be seen. The shale-prone facies grades shoreward into silt or sand-prone facies characterised by high continuity and high amplitude and basinward into a prograded slope

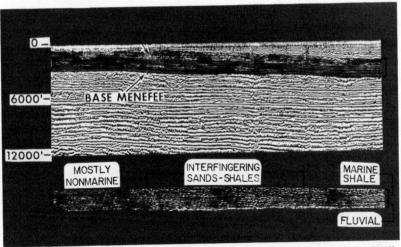

FIG. 9. Thirty-mile long portion of seismic section in the San Juan basin. Wells spaced about every mile provide subsurface control. The shaded portion of the section is repeated below. This unit is largely non-marine to the left and marine to the right. At the right side it is dominantly marine shale with a fluvial Menefee unit occupying the lower third. To the left this shale inter-fingers with sands, and only a thin portion of shale remains at the top of this unit at the left side. Reflections follow time-depositional surfaces through the marine and transition facies into non-marine facies. Although the continuity and reflections vary, the reflections generally are strongest in the inter-fingering transition zone. From Vail et al.[19] (Courtesy Exxon Production Research Company.)

facies. The sand-prone, low-amplitude facies grades landward into non-marine, low-continuity, variable-amplitude seismic facies and basinward into a high-continuity, high-amplitude marine facies. Low-continuity, variable-amplitude seismic facies often characterise fluvial deposition. Some of these patterns can be seen in Fig. 9.

The areal extent of seismic facies units is an important characteristic (Fig. 10). Especially important is whether units are widespread in sheet-like fashion or localised in fan shapes. Mounds of deposition are occasionally found, sometimes associated with turbidites.

An important aspect of seismic facies analysis, especially in virgin areas, is the identification of seismic sequences.[7] Seismic sequence boundaries are marked by one of three patterns above the sequence (down-lap, concordance or on-lap) and two or three patterns below the boundary (concordance, top-lap or truncation, the latter two sometimes being combined). These boundary situations are sketched in Fig. 11. The changes in reflection

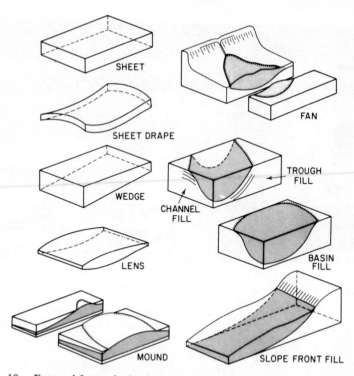

FIG. 10.   External form of seismic facies units. From Mitchum *et al.*[7] (Courtesy
Exxon Production Research Company.)

attitude which mark many sequence boundaries permit their fairly easy
recognition where the seismic data are of good quality. 'Concordance over
concordance' boundaries may be recognised by their continuity into places
where attitude differences make them evident, and they are often strong
reflectors, which aids their definition. A hiatus in deposition or erosion is
often associated with sequence boundaries. Depositional situation,
lithology or diagenesis often changes sharply at such boundaries, resulting
in appreciable change in rock properties and consequently a strong
reflection. An example of seismic sequence analysis is given in Fig. 12.

Vail and Todd[18] relate seismic sequence boundaries with major
worldwide eustatic variations. A pattern of gradual rise of sea level
followed by rapid fall of sea level has apparently occurred a number of
specific times throughout geological history (Fig. 13). This pattern appears
to be associated in an unspecified way with periods of inactive or active sea-
floor spreading. Because these eustatic variations are worldwide, seismic

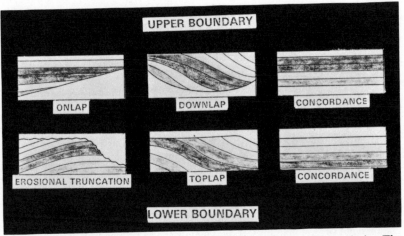

FIG. 11. Relationships strata may have with respect to sequence boundaries. The distinction between truncation, which is associated with an erosional surface, and top-lap is not always made. From Vail *et al.*[19] (Courtesy Exxon Production Research Company.)

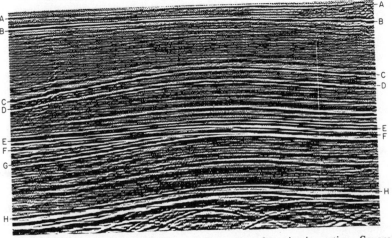

FIG. 12. Stratigraphic interpretation of a portion of a seismic section. Several reflectors have been lettered to facilitate reference. A was probably horizontal when deposited, so uplift of A indicates some structural growth subsequently. The interval from A to B thins on structure, indicating structural growth during deposition, but the amount of thinning differs on opposite sides of the crest. Whether the overall thickening to the right is local or regional could perhaps be

(continued)

sequences can be dated, if correlation with a master eustatic sequence chart can be established. Such correlation might be established by knowing the age at some point or recognising a distinctive feature of one of the cycles.

The distinctive features which sometimes provide the key are localised, deep-water fan sequences which indicate low sea-level periods. When the sea level is low, the continents tend to be emergent and sediments tend to be concentrated by marine canyons at the outer edges of the continental

answered if the line were longer. B is a strong reflector. There is on-lap on to B on both sides, but especially at the right, and truncation of data below B, especially at the left. B is an unconformity and had structural relief before the section above B was deposited. It might represent an 'on-lap over truncation' seismic sequence boundary, or it could also be explained as a result of local structure.

Relatively low velocities above C indicate clastic sediments with relatively little consolidation. The weak reflectors in the B to C interval indicate some lithology changes, but suggest that this section is mainly shale and mainly conformable. The relatively flat events are probably multiples which show up because the primary energy is so weak.

All intervals from C to E thicken to the right in a way which appears to have no relation to the structure. Perhaps the source of sediments was off to the right. There was probably no relief present during the deposition of this unit. D to E probably indicates one long cycle of rising sea level. The faint diffractions near the left end have no obvious explanation; perhaps their source is out of the plane. C and D are both strong events. The velocity increases sharply at C, indicating a markedly different depositional environment. C is probably a 'conformity over conformity' sequence boundary; D may or may not be a sequence boundary.

There is on-lap or down-lap of the sediments above E, but which is not clear because of subsequent structural growth. Combined with the thickening to the right, the evidence favours down-lap. Sea level fell at the time of E. E is probably a 'down-lap over conformity' sequence boundary.

Minor on-lap on to F from the left, and general thinning to the right, suggest no structural relief during this depositional period and conditions opposite to those of the section above E, i.e. a source of sediments to the left and rising sea level, F is clearly an angular unconformity and an 'on-lap over truncation' sequence boundary.

Fairly strong events from F to G suggest marked contrasts, possibly some high depositional energy. The units from F to H are reasonably thick, with constant thickness. H is another very strong event—another 'conformity over conformity' sequence boundary.

Below H the section is marked by many diffractions, indicating a surface of large massive blocks in a jumbled fashion—probably blocks of anhydrite or limestone and mobile salt. The salt probably did not move during H to F time, had a period of movement prior to the F unconformity, did not move during E to C time, perhaps not until nearly B, moved appreciably prior to, and during, the B unconformity, and had continued motion to post-A time. The axis of thinning seems to be moving to the left. (Courtesy Seiscom Delta Inc.)

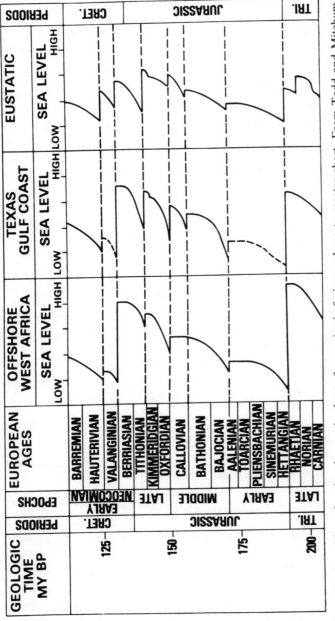

Fig. 13. Matching of regional eustatic charts from seismic facies analyses to master chart. From Todd and Mitchum.[17] (Courtesy Exxon Production Research Company.)

shelves, leading to localised fan sequences. On the other hand, when the sea level is high, the sea laps on to the continent and sedimentary patterns tend to be widespread.

VELOCITY ANALYSIS

The change in arrival time for a seismic reflection as the distance from the source to the detector ('offset') is varied depends upon the distribution of velocities within the Earth. The process of determining the distribution of velocities from such observations is called velocity analysis.[2,16]

While many algorithm variations are used to accomplish this objective, most arrange data systematically with respect to offset distance and then determine the quadratic curve which best fits each reflection event. Conceptually, this is equivalent to assuming a model in which the velocity is constant above each reflecting event, so that travel paths are straight lines. The agreement between the stacking velocity so determined and the actual velocity distribution in the Earth of course varies with the situation. The accuracy depends upon the offset distances, uncertain time delays near the source or geophones, the continuity and stand-out of reflections, noise, the dip of the reflecting interfaces, the direction of the seismic line with respect to strike and other factors. Generally speaking, for shot-to-geophone distances of the order of the depth of the reflecting event, the accuracy of velocity analysis is 1–2 % (perhaps 25 m/s).

The velocity in the Earth usually increases with depth in a reasonably systematic way, so that the Earth can be treated as a series of parallel layers, each of constant velocity, except in areas of marked lithological contrast or differential uplift. Stacking velocity values are equivalent to root-mean-square averages of the velocities, if the assumption of simple parallel velocity layering is close enough to the actual velocity distribution. Sometimes more complicated velocity distributions are assumed and velocity values are calculated in an iterative fashion, it usually being assumed that the distribution of velocities bears a systematic relationship to the bedding. Sometimes 'stripping' is used, that is the velocity to the first event is determined and then the positions of subsequent interfaces are determined one by one, the velocity being constant between interfaces.

The average velocity over the interval (interval velocity) between interfaces associated with different reflections can be determined for a given model of the velocity distribution. The calculation depends on the changes in arrival time and stacking velocity between the associated events. For reasonably thick intervals, that is for events separated by a half second or more where the velocity distribution is well behaved, the uncertainty in interval velocity may be only two or three times the uncertainty in the stacking velocities. However, as the interval becomes small, the uncertainty in interval velocity becomes large. It is usually not practicable to calculate interval velocity between events closer than a tenth of a second apart.

Interpretations of interval velocity calculations in terms of lithology or stratigraphy depend not only on the reliability of interval-velocity measurements, but also on the variety of lithologies which can have the same velocity. Both empirical studies of actual rocks and theoretical studies based upon a rock model are useful in determining how velocity depends on rock type, and both add insight to seismic interpretation.

MODEL FOR CLASTIC ROCKS

Most sedimentary rocks are heterogeneous, being made up of grains of different kinds packed together in different ways and with different degrees of cementation. The diversity makes it difficult to approach rock structure except in a statistical way. However, simple models of rock structure, especially of clastic rocks, are useful in predicting phenomena, especially velocity, and in enhancing our understanding, despite their simplicity compared with actual rocks.

Clastic rocks are composed of fragments of other rocks or of the shells of marine organisms. Such fragments settle gradually under the influence of gravity and slowly build up to form sedimentary rock units. Clastic rocks are identified by the size and distribution of the sizes of the particles of which they are composed, the mineral composition of those particles and the amount of cementation. Particle size depends on the population available for deposition and the sorting agencies which make the population of the sediments differ from that population. The distance from the source and the energy of the system which transports the particles are the dominant factors. Generally speaking, large particles are deposited first nearest the source, and the smaller particles are carried farthest and deposited most remotely from the source in calm water. Once laid down, sediments can be reworked, especially by wave action in the nearshore environment, which tends to sort particles by size and remove the finer ones. Particles are of nearly uniform size in some 'clean' rocks.

The simplest useful model is one of uniform size particles. Random packs of such particles have about 45% porosity when they begin to assume rigidity. Porosity depends on the geometry and not upon the size of the particles. Thus, packs of clay-size, sand-size, or soccer-ball size all have the same porosity. The porosity of a random pack varies slightly with particle shape, i.e. ellipsoids, cubes, rods, etc. The ability of fluid to flow through such a matrix depends on the size of the openings in the pack and so on the particle size. Thus, porosity and permeability are independent.

The porosity of a random pack will be less if the smaller particles fill some of the void space or if the particles deform at their points of contact because of pressure on the matrix. The consequence of the smaller particles filling void spaces is the familiar one of the lessened porosity of a dirty sand.

The stress on a matrix of particles tends to be concentrated at the points of contact between them. These adjacent particles will be in contact over an area rather than at a point, and the area will depend on the effective pressure. As the effective pressure increases, the contact area will increase and the effective elastic modulus will increase. Hence, the effective compressibility of the rock as a whole will depend upon the effective pressure, as well as on the elasticity of the minerals involved. The increase in elasticity with increase of effective pressure is the fundamental reason why seismic velocity increases with depth of burial. Cementation also increases the contact area between adjacent grains, and thus the effective elastic modulus and the velocity.

The effective stress on the matrix depends upon both the external pressure, which tends to squeeze the matrix, and the pressure of the fluid within the pore space, which tends to hold the matrix open. The picture which conforms with experimental evidence is that the effective stress is simply the difference between the overburden pressure and the fluid pressure. Both overburden and interstitial fluid pressures usually are nearly proportional to depth, so depth of burial can be thought of as a variable.

This model of a rock predicts that the seismic velocity will vary as the 1/6th power of the depth, a conclusion which experimental evidence generally supports.

The model we have been developing explains broadly the variation of porosity with depth as well as the variation of velocity. Since porosity is such a fundamental characteristic of a rock, porosity is often considered the dominant factor in determining seismic velocity.

Compaction under the influence of gravity alone is ordinarily a process of gradually increasing the effective stress, and therefore of systematically lowering the porosity and increasing the seismic velocity. In basins which have been continually subsiding under the weight of their own deposits and isostatic adjustment, the same systematic decrease of porosity and increase of seismic velocity with depth should be seen. This is in general agreement with observations. Compaction and velocity curves for such situations in all parts of the Earth are similar, and the same velocity–depth relation often is appropriate over large areas, despite differences in the attitude or age of the rocks. One usually does not have to worry about severe distortion because of lateral variations of velocity. The fact that compaction and seismic velocity depend mainly on depth greatly simplifies seismic interpretation in well-behaved situations, allowing one to develop a feel for structure despite the fact that a seismic section does not represent a cross-section through the Earth, even a distorted one. Where the sediments have been subjected to significant uplift and where stresses other than gravity may have been dominant, more complicated velocity and interpretation problems are expected.

The compaction of a rock under pressure produces irreversible changes in the matrix. Most of the pore space which is squeezed out of a rock by subjecting it to external pressure will not be regained once that pressure is removed, because the particles will have been permanently deformed. The rock, subsequently, may be able to withstand lesser pressures without further deformation. Thus, rocks retain an effect of the greatest pressure to which they have been subjected. Since porosity and seismic velocity both depend upon the maximum stress to which the rock has been subjected, and this maximum stress usually will have been when the rock was buried at its deepest point, seismic velocity has a component which 'remembers' the greatest depth of burial. This fact can sometimes be used to reconstruct the maximum depth of burial from seismic velocity measurements.

A decrease of porosity with depth of the burial implies that the water which formerly filled the void space has been removed from the system. Sometimes, however, sediments have been sealed so that the escape route for water is cut off. Under these circumstances, some of the overburden pressure will be transmitted to the fluid, and the fluid pressure will be greater than normal. Such abnormal pressure zones are found in the Earth, especially associated with shales. Increasing the fluid pressure is equivalent to decreasing the effective stress, so high-pressure zones are characterised by low seismic velocity. A rock under abnormal fluid pressure feels the same effective pressure as is appropriate to a shallower depth, and it has a velocity appropriate to that shallower depth.

Experimental data show that seismic velocity generally increases with the age of rocks. Almost anything which happens to a rock tends to decrease its porosity and increase its seismic velocity, such as stress over long periods of time, long-term flow or cementation. Since older rocks have had more time to be subjected to such processes, one expects a decrease in porosity and an increase in velocity with age.

DATA ON ROCK VELOCITY

Properties can be measured on samples of rocks taken from outcrops or boreholes, in boreholes by logging tools, and with seismic waves such as are used in seismic exploration. Measurements of these different kinds differ in one major regard: frequency. Seismic frequencies are in the 10–100 Hz

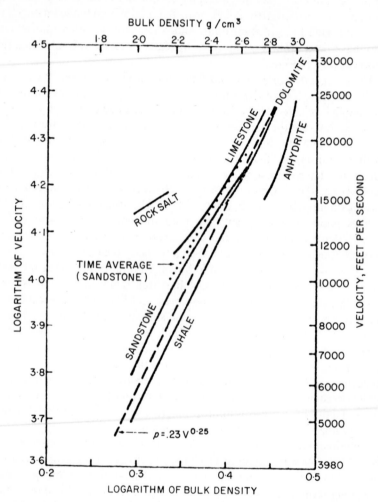

FIG. 14. Velocity v. density for rocks of various lithological types. From Gardner *et al.*[4]

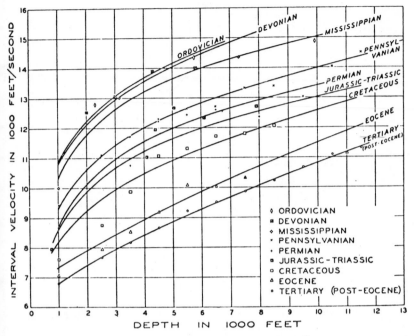

FIG. 15.   Velocity v. age. From Faust.[3]

range. logging tools operate in the ~20 kHz range, and measurements on samples are usually made at still higher frequencies (~MHz). Most data indicate that velocity does not vary appreciably over this wide range of frequencies.

The component rock minerals have seismic velocities significantly greater than the velocities observed for rocks. The seismic velocity of the minerals can be determined with reasonably high precision and little scatter, whereas the velocity of rocks of the same general class varies considerably.

The relationship between velocity and lithology, density or porosity, and age is illustrated in Figs. 14 and 15. Variation with density or porosity amounts to essentially the same thing, since, for water-saturated sediments, knowing the value of one determines the other. The velocity of Gulf Coast sands and shales as a function of depth is shown in Fig. 16.

Unfortunately the mass of data on velocity measurements has so much scatter that only very general conclusions can be drawn. Scatter in data usually means other variables are involved and that they have not been

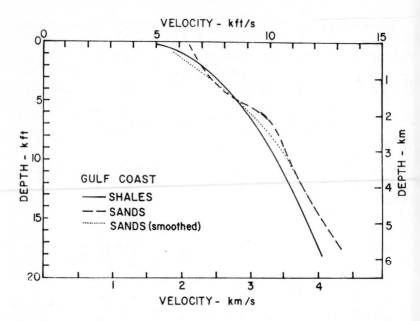

FIG. 16.   Velocity v. depth for sands and shales in the US Gulf Coast, consolidated only by the effect of the overburden pressure. The dotted curve has been smoothed to produce a more systematic curve. Data from Gregory.[5]

taken into account. One can do a better job of predicting from velocity data if local data on the dependence of velocity on specific factors are available.

Figure 14 shows the dependence of velocity on sediment lithology, but the overlap of types is so large that predictions are useful only in the extremes. Generally, low velocities indicate clastic sand and/or shale sequences, and high velocities indicate carbonates or evaporites. Sometimes sand/shale ratios have been predicted from velocity measurements, on the theory that sand velocities are sufficiently larger than shale velocities to be significant. Data in some areas support this point of view, while data in other areas contradict it.

In clastic environments abnormal fluid pressure often can be predicted where primary reflections can be obtained within or below the abnormal pressure zone. Sometimes the degree of abnormal pressure can be predicted with fair reliability.[8,9] Some literature has claimed to predict geothermal gradient, but other literature disputes the method. The interpretation of velocity anomalies in detecting gas accumulations is discussed in Chapter 9.

While there is cause to be cautious in making porosity and lithological interpretations from velocity data, significant improvements in measurement techniques are under way. Some companies derive important geological conclusions from velocity measurements, although the facts have not reached the literature.

OTHER SEISMIC MEASUREMENTS

Unsupported and confusing claims of interpretation are made based on measurements of the attenuation of seismic energy. Attenuation over the seismic range increases linearly with frequency, and appears to depend significantly on fluid content. However, techniques for the measurement and interpretation of attenuation data are not generally available.

Measurements using shear waves suggest a number of ways of deriving rock properties, such as Poisson's ratio and fluid content, but measurement and interpretation techniques are not yet available. Considerable research is going on in this area, both by industry and for earthquake prediction, so we may hope for new techniques in the future.

CONCLUSIONS

Events in seismic data do not just happen; they are caused. Several recently developed techniques open new interpretation possibilities. As data improve in quality and our understanding of rock properties increases, we may reasonably expect to be able to deduce more rock properties with higher precision and answer more of our questions about the nature of the Earth.

REFERENCES

1. BROWN, L. F. and FISHER, W. L. (1976). Seismic facies reflection patterns: examples from Brazilian rift and pull-apart basins; from AAPG–SEG school on 'Stratigraphic Interpretation of Seismic Data'.
2. COOK, E. E. and TANER, M. T. (1969). Velocity spectra and their use in stratigraphic and lithologic differentiation, *Geophysical Prospecting*, **17**, 443–8.
3. FAUST, L. Y. (1951). Seismic velocity as a function of depth and geologic time, *Geophysics*, **16**, 192–206.
4. GARDNER, G. H. F., GARDNER, L. W. and GREGORY, A. R. (1974). Formation

velocity and density, the diagnostic basics for stratigraphic traps, *Geophysics*, **39**, 770–80.
5. GREGORY, A. R. (1976). Some aspects of rock physics from laboratory and log data that are important to seismic interpretation; from AAPG–SEG school on 'Stratigraphic Interpretation of Seismic Data'.
6. LAING, W. E. (1977). Case histories and field documentation of seismic reflection patterns, in: *Stratigraphic Interpretation of Seismic Data*, Am. Assoc. Petrol. Geologists, Tulsa.
7. MITCHUM, R. M., VAIL, P. R., SANGREE, J. B. and THOMPSON, S. (1976). Stratigraphic interpretation of seismic reflection patterns in depositional sequences; from AAPG–SEG school on 'Stratigraphic Interpretation of Seismic Data'.
8. PENNEBAKER, E. S. (1968). An engineering interpretation of seismic data; Society of Petroleum Engineers Preprint No. 2165.
9. REYNOLDS, E. B. (1970). Predicting over-pressured zones with seismic data, *World Oil*, **171**(5), 78–82.
10. SANGREE, J. B., FRAZIER, D. E., WAYLETT, D. G. and FENNESSY, W. J. (1976). Recognition of continental-slope seismic facies, offshore Texas–Louisiana; from AAPG–SEG school on 'Stratigraphic Interpretation of Seismic Data'.
11. SANGREE, J. B. and WIDMIER, J. M. (1974). Interpretation of depositional facies from seismic data. Paper presented at *Symposium on Contemporary Geophysical Interpretation*, Geophysical Society of Houston, 4–5 Dec., 1974.
12. SHERIFF, R. E. (1976). Inferring stratigraphy from seismic data, *Am. Assoc. Petrol. Geologists Bull.*, **60**, 528–42.
13. SHERIFF, R. E. and FARRELL, J. (1976). Display parameters of marine geophysical data; Offshore Technology Conference, OTC Paper 2567, Dallas.
14. SHERIFF, R. E. (1977). Limitations on the resolution of seismic reflections and the geological detail derivable from them, in: *Stratigraphic Interpretation of Seismic Data*, Am. Assoc. Petrol. Geologists, Tulsa.
15. TANER, M. T. and SHERIFF, R. E. (1977). Application of amplitude, frequency and other parameters to stratigraphic and hydrocarbon determination, in: *Stratigraphic Interpretation of Seismic Data*, Am. Assoc. Petrol. Geologists, Tulsa.
16. TANER, M. T. and KOEHLER, F. (1969). Velocity spectra—digital computer derivation and application of velocity functions, *Geophysics*, **34**, 859–81.
17. TODD, R. G. and MITCHUM, R. M. (1976). Identification of Upper Triassic, Jurassic and Lower Cretaceous seismic sequences in offshore West Africa and Gulf of Mexico; from AAPG–SEG school on 'Stratigraphic Interpretation of Seismic Data'.
18. VAIL, P. R. and TODD, R. G. (1976). Interpreting stratigraphy and eustatic cycles from seismic reflection patterns; from AAPG–SEG school on 'Stratigraphic Interpretation of Seismic Data'.
19. VAIL, P. R., TODD, R. G. and SANGREE, J. R. (1976). Chronostratigraphic significance of seismic reflections; from AAPG–SEG school on 'Stratigraphic Interpretation of Seismic Data'.

Chapter 9

'BRIGHT SPOT' TECHNIQUES

CHARLES B. STONE

Stone Geophysical Company, Conroe, Texas, USA

SUMMARY

'Bright spots' were discovered in 1968 and have been in common use in the petroleum industry since 1972, resulting in numerous discoveries of gas and dry holes. The reliability of the 'bright spot' method must be questioned due to the dry holes. It is the purpose of this paper to discuss the reasons for the dry holes, place the blame where it belongs, and recommend procedures to reduce dry hole risk.

'Bright spot' techniques require severe over-simplification of the complex process of seismic wave propagation, which leads the user to concentrate upon the relationship of high seismic reflection amplitudes to hydrocarbons while neglecting the more numerous other causes for high amplitudes. Displaying true seismic amplitudes is an inexact science which can lead to erroneous conclusions.

A statistical use of other geophysical hydrocarbon indicators is presented along with a recommended future procedure for analysing variations in the individual components of the common-depth-point stack.

It is concluded that failures are due to over-simplification, inadequate understanding and failure to adequately assess risks in using the methods. Despite the misuses, successes are encouraging enough to predict a 'bright' future for geophysical hydrocarbon indicators as a valuable tool for the explorationist.

275

INTRODUCTION

'Bright spot' exploration techniques are fairly recent developments in geophysics. Their unusual name is derived from the fact that high-amplitude reflections appear 'brighter' than low-amplitude reflections when observed on a seismic profile. This 'brightness' is important because

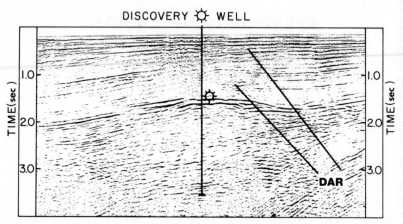

FIG. 1.   A 'bright spot' at a gas reservoir. DAR: Detection of Amplitude Relationships. (Petty-Ray Geophysical Division, Geosource, Inc.)

high-amplitude events may indicate the presence of hydrocarbons. The name has been rather loosely applied, and is now commonly accepted by the petroleum industry to represent any geophysical measuring technique which is used to predict the existence of hydrocarbons prior to drilling. Figure 1 illustrates the appearance of a typical 'bright spot' reflection event.

HISTORY

Attentive geophysicists working for two of the major oil companies operating in the Gulf of Mexico in 1968, discovered a relationship between high-amplitude seismic reflections and reservoirs containing gas. Tight security measures were imposed upon the discovery and it was not until 1972 that information about this 'bright spot' relationship was leaked to the remainder of the industry. The discovery arrested an alarming decline in geophysical exploration and created a minor boom in the geophysical industry. A number of technical papers were published on 'bright spots', and it became common practice to hold 'bright spot' symposia internally within individual companies and externally through co-operating geophysical societies. Fortunately, digital instrumentation and computer technology had independently advanced more rapidly than interpretation technology, and were timely available to implement the needs of the new interpretation method. Unfortunately, however, an atmosphere of glamour accompanied the news that a direct oil finder was now a reality. Consequently, geophysical contractors overemphasised the importance of the new method, and oil company management overreacted by placing more confidence in the method than was warranted. Initial successes based on drilling 'bright spot' anomalies were interspersed with failures, and it soon became apparent that 'bright spots' alone did not always indicate the presence of hydrocarbons. Despite some adverse public comments by representatives of companies who unsuccessfully used 'bright spots', no serious backlash against the method developed, because the more competent companies recognised its importance for what it was; another valuable tool in the hands of the geophysicist. The importance of the discovery of 'bright spots' should not be underestimated, for the discovery produced an entirely new approach in geophysics, forced the geophysicist and management to look more closely at their seismic data, arrested the decline in geophysical exploration and accelerated geophysical research in the direction towards finding new methods to decrease the risk of drilling dry holes.

BASIC PRINCIPLES

Velocities and densities of sedimentary rocks vary according to their mineral composition, degree of compaction, porosity, pressure and fluid/gas content contained within the pores of the rock matrix. It is the contrast between velocities and/or densities across an interface between two rocks which causes seismic energy to be reflected from the interface. A convenient form of expressing the velocity–density relationship of an individual rock is by the mathematical product of its density and velocity, a property known as its acoustic impedance. The formula for acoustic impedance is:

$$I_a = DV \tag{1}$$

where I_a is the acoustic impedance, D is the density and V is the velocity.

Knowing the amplitude of a signal impinging upon an interface, and knowing the acoustic impedances on either side of the interface, provides enough information to predict the amplitude of a signal which will be reflected from the interface. The procedure by which this is accomplished is through the substitution of the acoustic impedances into a formula for the reflection coefficient. The output of this formula is a ratio of the amplitude of the reflected signal to the amplitude of the impinging signal. For example, a reflection coefficient of 0·5 would indicate that the reflected signal was half (50%) as large as the signal which struck the interface. In actuality, a reflection coefficient of 0·5 would be unusual; coefficients of 0·1 are common, but coefficients of less than 0·1 occur most frequently.[1] Thus the reflected amplitudes are commonly less than 10% of those of the impinging signal. The remaining 90% is transmitted through to deeper horizons which in turn reflect and transmit a progressively decreasing impinging signal and results in ever-decreasing reflection amplitudes with depth.

The formula for the reflection coefficient is:

$$\frac{A_r}{A_i} = R_c = \frac{D_2 V_2 - D_1 V_1}{D_2 V_2 + D_1 V_1} \tag{2}$$

where A_r is the amplitude of the reflected signal, A_i is the amplitude of the impinging signal, R_c is the reflection coefficient, and $D_2 V_2$ and $D_1 V_1$ are the acoustic impedances of the deeper and shallower layer respectively.[2] Note that the output of the formula may be either positive or negative depending upon which layer has the larger acoustic impedance. A positive value indicates that the deeper layer has the larger impedance and the interface reflects a compressional-type signal, whereas a negative value indicates that the shallower layer has the larger acoustic impedance and the interface reflects a rarefaction-type signal. This reaction may be considered as representing the polarity of the signal, and by simple analogy a pulse on a seismic trace breaking upward would be positive, and one breaking downward would be negative.

It has been shown that if the impinging signal amplitude and the velocities and densities are known, the reflection amplitude can be calculated. In actual practice, the reverse approach is used; reflection amplitudes are displayed and the amplitudes are used to draw conclusions about the densities and velocities which produced the changes. If an assumption is made that all parameters affecting densities and velocities across a reflecting interface remain *constant*, except for those due to the gas/fluid contents of the rocks, then any variation in observed amplitudes of reflected events can be attributed to a change in pore contents. It is this simple principle of relating observed changes of relative amplitudes to changes of velocity and density caused by changes in pore contents which serves as the basic background for 'bright spot' technology. The standard procedure is to scrutinise particular seismic reflections on a seismic profile for lateral variations in relative amplitude, and to assume that any pronounced amplitude changes of reflections coming from any suspected reservoir are directly related to hydrocarbons in the reservoir. Any observed high amplitudes are called 'bright spots'.

A simplistic preparation of the basic background on the 'bright spot' method has now been completed. An understanding of the few simple principles making up this background will suffice to allow an explorationist to draw conclusions about the potential existence of hydrocarbons based on observation of amplitudes on a 'so-called' true-amplitude seismic profile. In fact, the majority of past decisions have been made in just this manner. This accounts not only for the successes but also for the failures. For what has actually occurred both herein and in actual circumstances is that a procedure has been developed and used which is based upon a highly complex subject which has been severely oversimplified. This oversimplification is the 'Achilles Heel' of the method. Explorationists have become 'bright spot' interpreters, because 'bright spots' are easy to see, and if we make the arbitrary assumption that they indicate hydrocarbons, almost anyone can become an 'instant expert'. A second look at Fig. 1 will confirm the ease with which a 'bright' spot can be seen to indicate gas in a reservoir.

PITFALLS

Under basic principles, the deceptive simplicity of the theory of 'bright spots' has been presented; now we will look at some valid reasons for counteracting this misleading simplicity.

It should be obvious that the basic assumption of relating changes in velocities and densities only to changes in pore contents ignores the other common causes for such occurrences. Changes in lithology, tuning effects of varying bed thicknesses, focusing of seismic waves, enhancement by multiples arriving in phase, variations in the shot and recording mediums causing changes in outgoing and incoming signals, the unpredictable filtering effect of the Earth, recording instrument parameters, and of equal importance, the selection of processing parameters by the data processing

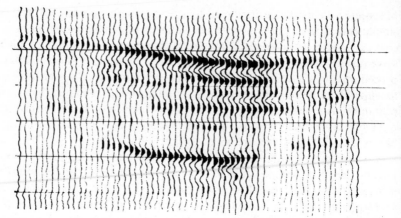

FIG. 2.   Focusing of energy reflected from a sharp syncline. (Transocean Oil, Inc.)

centre, are all potential contributors to amplitude changes. Figure 2 illustrates the focusing effect of a sharp syncline, Fig. 3 the concentration of energy at a diffracting point and Fig. 4 the effect of a lithological change.

Some comments are also necessary about the formula for calculating reflection coefficients. First, predicting the true amplitude of a reflected signal requires knowing the true amplitude of the impinging signal. In actual practice, the impinging signal amplitude is never accurately known. Measurements of outgoing seismic signals are seldom made because of existing detrimental near-field effects, and even if known, the diminishing effects on the transmitted signal due to the reflection and transmission at all shallower reflection boundaries would have to be known, in order to calculate the signal available to impinge upon deeper layers. Thus, to predict true impinging signal amplitudes would require that we know not only the outgoing signal, but all of the physical parameters of all of the

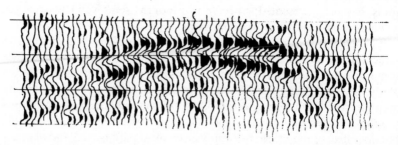

FIG. 3.   Concentration of energy at a diffracting point. (Transocean Oil, Inc.)

reflecting horizons. This is an impossibility because it presupposes that we know what we are really trying to determine: the physical parameters of the reflection horizons. Second, the formula for reflection coefficient is valid only for events striking an interface at normal incidence (perpendicular to the interface). Seismic profiles are commonly considered as vertical sections created by shots and detectors placed at normal incidence. This practice is

FIG. 4. 'Bright Spot' caused by a lithological change.

warranted for simplification purposes, but the interpreter of the common-depth-point (CDP) profile, which is normally used for 'bright spot' analyses, should not neglect the fact that the amplitudes on this profile are composites of individual amplitudes which were not obtained at normal incidence. Figure 5 illustrates this fact.

Maybe less obvious is our lack of a clear understanding of the one item we are utilising from the seismic profile: amplitude. What really is the factor, amplitude? And what operations have been performed upon it both by the Earth and by processing geophysicists prior to displaying it for use on a seismic profile? To be a competent interpreter of 'bright spots' requires a knowledge of the entire seismic reflection system.

FIG. 5. Comparison of normal incidence point and CDP.

It would be fortuitous indeed if the amplitude of seismic events depended only upon the physical changes at the reservoir. Unfortunately, seismic waves must be generated at the surface of the Earth, travel through the Earth to the reservoir where they are reflected back to the surface and recorded on sensitive seismic instruments. The Earth is unpredictable; it absorbs signals, disperses energy randomly, distorts signal amplitudes and helps contribute noise which further contaminates recorded signals.

Consideration is now necessary of three important factors which occur in the field, but which are compensated for in seismic data processing centres: spherical divergence, instrumental amplification and inelastic attenuation.

Spherical Divergence

If we assume a homogeneous Earth and a constant velocity of travel, we can consider seismic energy as starting at a point at the surface and expanding down into the Earth as a sphere of ever-increasing radius. The energy which was previously concentrated at the point must at any later time be spread over the entire surface of the expanded sphere. The depth of penetration into the Earth will be the radius of the sphere calculated by multiplying the time of expansion by the constant velocity. It should be obvious that a penetration of the order of kilometres will produce a very large sphere, and hence the energy remaining on a small unit area on the surface of the sphere, which is available to come into contact with a reflecting point, is very small. Amplitude is related to energy by the fact that energy is the square of the amplitude. Thus, we are dealing with a very small amount of energy at the reflecting point, which will be further reduced by taking its square root to get the amplitude, which in turn is additionally decreased by a reflection coefficient of less than one tenth, resulting in a very diminished reflected signal amplitude.

In reality the Earth is not homogeneous, nor is the velocity constant, and the assumed sphere is not a sphere; however, the fact must be faced that the amplitude is being physically diminished, and assumptions must be made to correct for this effect if we are to establish usable final amplitudes.

The area of a sphere, s, is a function of the square of its radius, r, multiplied by a constant:

$$s = kr^2 \tag{3}$$

As the sphere expands, the available energy is spread over the entire surface of the sphere and the energy, E, on each unit area of the sphere is reduced by the inverse value:

$$E = \frac{1}{kr^2} \tag{4}$$

If we neglect the constant, the energy per unit area becomes:

$$E = \frac{1}{r^2} \tag{5}$$

Energy is equal to the square of the amplitude, A:

$$E = A^2 \qquad (6)$$

Equating eqns. (5) and (6):

$$A^2 = \frac{1}{r^2} \qquad (7)$$

resulting in an equation stating that amplitude varies inversely as the radius:

$$A = \frac{1}{r} \qquad (8)$$

Because the radius is equal to the length of travel path in the Earth, we can compute the radius in terms of time expansion multiplied by the velocity of expansion:

$$r = Vt \qquad (9)$$

with V being velocity and t being time. Substituting for r in eqn. (8) we arrive at a final equation for amplitude at times of expansion:

$$A = \frac{1}{Vt} \qquad (10)$$

This is the equation for a correction for 'spherical divergence', and suffers somewhat from our assumptions, but none the less is commonly applied in deriving 'bright spot' amplitudes. t is known accurately and V is available or calculated with reasonable accuracy.

Instrumental Amplitudes
The development of binary gain amplifiers made 'bright spot' recording possible. The recording system is required to receive very small signals, amplify them on the order of one million times, store the amplification factor (gain) applied to the signal on tape for future use, and perform these functions with minimal distortion of the signal. To accomplish this requires high precision equipment. The instrument manufacturers have performed nobly and produced instruments of such high quality that corrections for amplification (gain restoration) are accurate and of no concern to geologists, if handled capably by field and data processing centre employees.

Inelastic Attenuation

The effect of the Earth on signal amplitudes cannot be predicted or calculated by any known equation. This problem must be approached empirically. We have already seen that amplitudes decrease due to spherical expansion, but in addition they suffer losses from the Earth itself due to absorption and dispersion. By using the signal recorded on tape we can plot the reflection amplitudes as a function of time. This function must then be reduced by reversing out the amplification of the instruments (gain restoration), and by applying the correction to compensate for spherical divergence. The remaining amplitudes plotted as a function of time *after* correction for divergence and amplification can serve as an empirical guide to the rate of decay of energy due to the Earth's effect. By constructing a series of curves for an area, an *average* curve can be established showing the average energy decay in the area.

Use of this curve allows the derivation of a regression coefficient representing the rate of observed decay of energy, and an equation for expressing a correction due to 'inelastic attenuation' which is commonly in the form of decay with time:

$$A = A_0 \exp(-dt) \qquad (11)$$

with A the amplitude, A_0 the initial amplitude, t the time and d the regression coefficient determined empirically for the given area. The data processor removes the amplification effect of the field instruments, makes an estimated correction to compensate for the decrease in energy due to the expanding sphere, and applies the estimated correction for the 'inelastic attenuation' of energy in the Earth. Further selection of parameters for elimination of reverberations, selection of velocities for CDP compositing, and selection of final profile display parameters, all affect the final appearance of the 'bright spot' profile. Thus, 'bright spots' are merely representations of minute amplitude changes *after* corrections are applied for much larger effects which are based partly on recorded information, but which are largely based on necessary assumptions, accurate calculations and educated estimates by data-processing personnel. The interpretation geophysicist often finds himself restricted by some of the decisions made previously in the data processing centre.

It is not the intent of this chapter to dwell upon this procedure for amplitude reduction, for this is capably provided by geophysical data processing centres. It is, however, the intent to warn the user of 'bright spots' that he is dealing with a complex problem of looking at a final set of amplitudes which could be adequately described at best as relative amplitudes. As we have seen, Earth, instruments and man manipulate the amplitude non-uniformly based on unknown physical changes and necessary assumptions. So it should not be surprising that 'bright spot' methods often fail. For this reason, a simplistic approach to using amplitudes can be somewhat justified if the explorationist keeps firmly in mind the risk factors inherent in the over-simplification. The trend in the industry at present, however, is to assign some quantitative values as to the size of reservoirs based on observed amplitudes. This type of procedure is critically inexact, must be used with caution, and cannot tolerate any over-simplification.

OTHER GEOPHYSICAL HYDROCARBON INDICATORS

The discovery of 'bright spots' accelerated efforts by geophysicists to analyse more closely other events which are commonly observed on seismic profiles. As a result, several other geophysical hydrocarbon indicators have been categorised: polarity inversions, time sags, absorption of high frequencies, flat spot reflections, velocity slow-downs, dim spots,

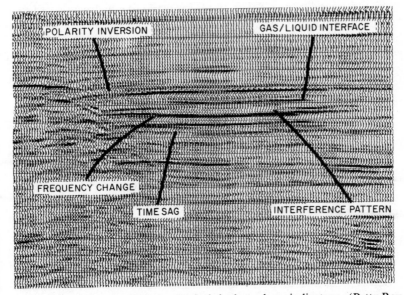

FIG. 6. Examples of other geophysical hydrocarbon indicators. (Petty-Ray Geophysical Division, Geosource, Inc.)

DISCOVERY ☼ WELL

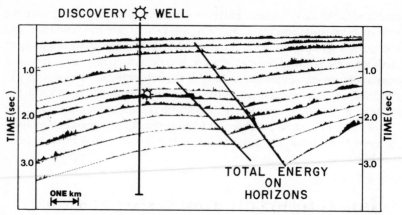

FIG. 7.   High levels of reflected energy from gas reservoir. (Petty-Ray Geophysical
Division, Geosource, Inc).

diffraction patterns at the edges of reservoirs, and constructive and
destructive interference patterns.[3] Figure 6 illustrates some actual
examples. Thus, 'bright spots' (high amplitudes) are just one form of a
much larger group of events which are classified as geophysical
hydrocarbon indicators. The methods for predicting hydrocarbons which
appear to have most merit, are those which stress the statistical advantage
of using all geophysical hydrocarbon indicators which are available, and
which assign only an equal factor of importance to measurements of

DISCOVERY ☼ WELL

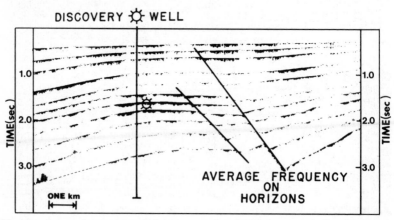

FIG. 8.   Lower frequency reflections associated with gas reservoirs. (Petty-Ray
Geophysical Division, Geosource, Inc.)

amplitude. Figure 7 displays an analysis of the seismic profile represented in Fig. 1, and shows high levels of reflection energy caused by gas in the reservoir; high energy levels are identified by the increased size of the dark areas above the lines representing the reflections. Figure 8 illustrates a shift to lower frequencies in the vicinity of the reservoir, caused by more rapid dispersion of high frequencies due to gas in the reservoir; increased low

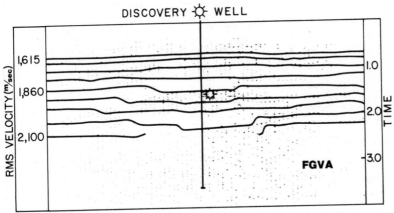

FIG. 9. Lowering of RMS velocities in and below a gas reservoir. FGVA: Fine Grain Velocity Analysis. (Petty-Ray Geophysical Division, Geosource, Inc.)

frequency content is shown by larger dark areas extending below the lines representing the reflection. Figure 9 shows by contour values the decreased RMS velocities in and beneath the reservoir caused by the low velocity of gas in the reservoir. These figures illustrate the confirmation of gas in the reservoir by measurement of three separate quantities (energy, frequency and velocity) thus statistically supporting the 'bright spot' which is apparent on the seismic profile of Fig. 1.

GENERAL DISCUSSION

Geophysical hydrocarbon location techniques have been used most successfully in marine exploration. This is because both sources and detectors are suspended in the uniform water medium which essentially does not vary from point to point. In contrast, the sources and geophones are placed in a non-uniform surface medium in land exploration, and the non-uniformity contributes unpredictable variations in both outgoing and

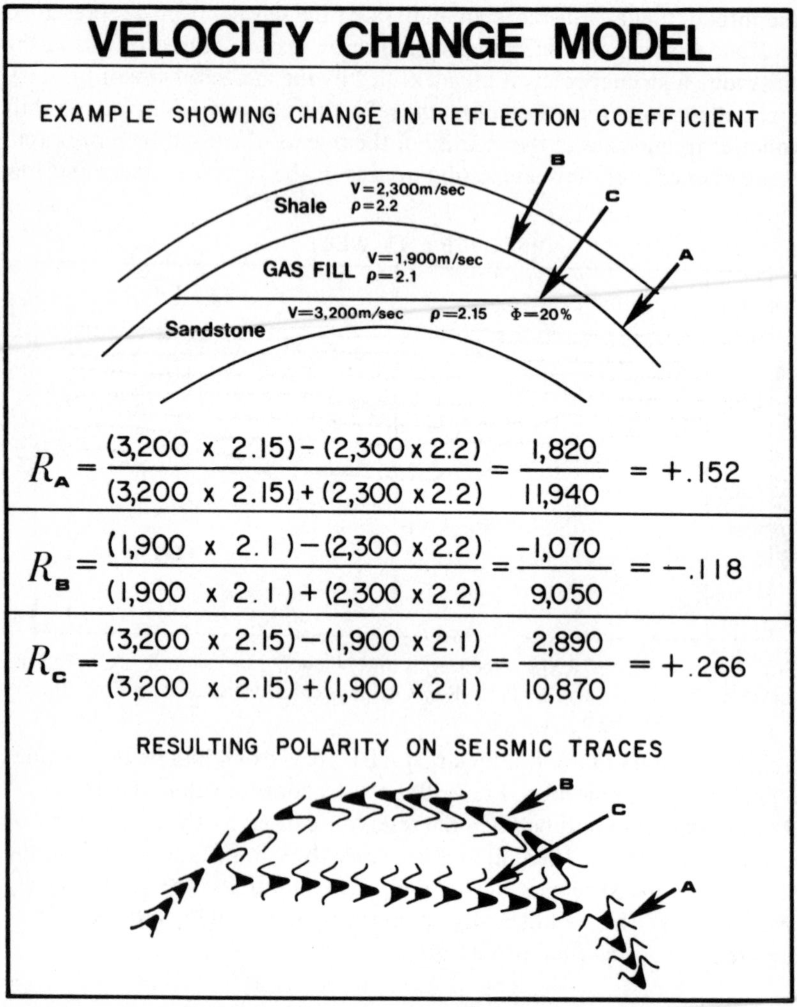

VELOCITY CHANGE MODEL

EXAMPLE SHOWING CHANGE IN REFLECTION COEFFICIENT

Shale V=2,300m/sec ρ=2.2

GAS FILL V=1,900m/sec ρ=2.1

Sandstone V=3,200m/sec ρ=2.15 Φ=20%

$$R_A = \frac{(3,200 \times 2.15) - (2,300 \times 2.2)}{(3,200 \times 2.15) + (2,300 \times 2.2)} = \frac{1,820}{11,940} = +.152$$

$$R_B = \frac{(1,900 \times 2.1) - (2,300 \times 2.2)}{(1,900 \times 2.1) + (2,300 \times 2.2)} = \frac{-1,070}{9,050} = -.118$$

$$R_C = \frac{(3,200 \times 2.15) - (1,900 \times 2.1)}{(3,200 \times 2.15) + (1,900 \times 2.1)} = \frac{2,890}{10,870} = +.266$$

RESULTING POLARITY ON SEISMIC TRACES

FIG. 10. Effect of gas fill on reflection coefficients. (Petty-Ray Geophysical Division, Geosource, Inc.)

incoming signals. Thus, land applications present a challenge to research organisations to develop adequate techniques to overcome and/or measure detrimental surface signal effects.

Geophysical hydrocarbon indicators are also more oriented toward gas exploration. The velocities and densities of gas vary radically from those of the formation waters which they displace in a reservoir, thus creating substantial contrasts which produce observable changes. Figure 10 illustrates with a model, the effect of gas on reflection coefficients and expected reflection patterns. Contrary to this, petroleum has essentially the same seismic velocities and densities as water, being slightly slower and slightly lighter, and thus creates no appreciable contrasts when displacing the formation waters and produces little evidence to use as hydrocarbon indicators.

The 'bright spot' nomenclature has been slow to change to a more apt description of these events as 'geophysical hydrocarbon indicators', and more realistically one should now consider a modification to 'geophysical gas indicators'.

DIRECTIONS FOR THE FUTURE

Almost all of the current 'bright spot' techniques use the composited traces of the common-depth-point (CDP) section as the basis of analysis. The CDP seismic section is necessary for elimination of multiple reflections so that 'bright spots' can be observed. Once the 'bright spot' has been identified, a decision must be made as to the feasibility of further derivation of quantitative predictions about the areal size and content of the reservoirs. This is a complex process and analysis of the CDP section is too gross an approach to provide any reasonable quantitative figures. The CDP stack suffers from its summation of traces which have passed by different paths through the Earth, with each path contributing its own filtering, attenuation, absorption, velocity, amplitude, signal and other physical effects. Figure 11 shows paths which do not all pass through a gas reservoir. When the traces are summed and the summation analysed, the individuality of each of the traces is lost. The most logical approach then, is

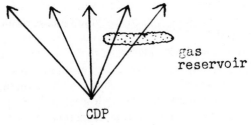

FIG. 11.   Some CDP trajectories missing a reservoir.

to take apart the CDP stack after it has served as the means to identify the 'bright spot', and then analyse the individual traces to isolate effects due to the presence of any hydrocarbons. Care must be taken to recognise the contribution of multiple reflections which will be retained in the individual traces. It is suggested that each individual trace be sorted and displayed not only in CDP gathers, but in gathers of common shot, common receiver and common distances. Analyses should be made of these common gathers to derive an average effect of the common parameter, and then to search the individual traces for deviations from this average. An intense effort to relate deviations to their most likely causes appears to be a more sound approach than to continue to attempt to analyse the composited CDP traces.

It will be apparent to geophysicists involved in data processing that this will be a tremendous task, requiring the processing, display and analysis of an increase of possibly a hundredfold in data handling.

The decision to conduct such analyses will be an economic one which will authorise the necessary funding to provide the additional data processing personnel, computer time, and skilled geophysical interpreters which will be required to provide quantitative predictions. Viewed in relation to the rapidly increasing costs of leasing, drilling, production, pipe-lining, refining and marketing, an increase in geophysical expenditures which will make predictions more accurate is one of the best available bargains in the petroleum industry.

CONCLUSION

'Bright spot' technology is in the early stages of development. It has reversed a trend of declining geophysical prominence, and hence its discovery will go down in history as one of the most influential advances in geophysics. The inception occurred at a time when instrumentation and computer technology were available to capitalise upon its use. Possibly its greatest ultimate benefit will be derived from pointing out that interpretation techniques had lagged far behind that of the instrumentation and computer capabilities, and its discovery accelerated the development of new interpretational methods. The art of recognition of 'bright spots' and other potential hydrocarbon indicators can be used in its simplistic form if risks are clearly considered. Future emphasis will be upon the development of techniques to analyse geophysical hydrocarbon indicators in order to predict reservoir qualities and quantities. Hopefully, it has been shown that such predictions are a risky business. The author knows of no way to assign quantitative values accurately to reservoir contents based on geophysical hydrocarbon indicators. At present, the systems of most merit appear to be those used by progressive oil companies in cataloguing sets of empirical examples of 'geophysical hydrocarbon indicators' as compared to actual reservoir data. Observed seismic anomalies are then used to search through the catalogue of anomalies to find an analogy to a catalogued anomaly, and the same reservoir quantities are then assigned to the observed anomaly. This procedure is, of course, possible only for oil companies which have access to the reservoir measurements, and thus geophysical contractors, who find it almost impossible to obtain such data because of its competitive value, are excluded.

In summary, it is clear that the use of 'bright spot' techniques is not an exact science. Considerable improvements in both data processing techniques and interpretation will be required to improve its record of predictions.

In the oil business it is not unusual to hear a statement that 'no oil will ever be found in this geographic area', and yet years later to find the area has become a major producer. Statements about 'bright spot' techniques fall somewhat into this category. Currently, it appears that no geophysical technique will ever be capable of predicting the existence of hydrocarbons with a high degree of accuracy prior to drilling. But are we not fortunate that those attentive geophysicists working in the Gulf of Mexico in 1968 did not know any better?

REFERENCES

1. O'DOHERTY, R. F. and ANSTEY, N. A. (1971). Reflections on amplitude, *Geophysical Prospecting*, **19**, 430–50.
2. CRAFT, C. (1973). Detecting hydrocarbons—for years the goal of exploration geophysics, *Oil and Gas J.*, **71**(8), 122–5.
3. STONE, C. B. (1975). *Exploration and Economics of the Petroleum Industry*, Matthew Bender and Co., New York, 13.

Chapter 10

SHALE DENSITY STUDIES AND THEIR APPLICATION

WALTER H. FERTL

Dresser Industries Inc., Houston, Texas, USA

SUMMARY

Shale density values can be measured on rock samples at the surface or recorded in boreholes by means of geophysical well logging operations. Such shale density data can be related to several important geologic phenomena including location within a sedimentary basin, unconformities, geologic age, structural configurations, lithologic variations and subsurface formation pressures in excess of hydrostatic.

Recognition and quantitative evaluation of over-pressures, and computation of formation fracture pressure gradients from shale density variations versus depth are used by the petroleum industry to drill boreholes with maximum safety and minimum cost.

In formation evaluation the shale density assists in determining the effective pore space in reservoir rocks containing hydrocarbon resources or fresh water supplies. In addition, the four basic elastic rock constants, Young's modulus, bulk modulus, shear modulus and Poisson's ratio, can be calculated from density and acoustic data. These elastic constants are widely used in petroleum engineering.

INTRODUCTION

Definition and Composition of Argillaceous Sediments

Clay is used as a rock term and also as a *particle size term*. Generally speaking the term clay implies a natural, earthy, fine-grained material which develops plasticity when mixed with a small amount of water.[28] Different disciplines employ clay as a particle size term differently.

Geologists tend to follow the Wentworth scale, which classifies material finer than 4 microns† as clay grade. Soil scientists, however, characterise any material smaller than 2 microns as clay grade.

Clay minerals which are important components of argillaceous sediments, owe their origin to three principal processes:

(1)   Detrital inheritance from pre-existing rocks and sediments;
(2)   Transformation of the clays from one form to another in the sedimentary environment owing to their instability at low temperatures and in various ranges of pH and Eh-potentials;
(3)   Neo-formation of clay minerals *in situ*.[58]

Clays may contain varying percentages of clay-grade material consisting of both clay-mineral and non-clay-mineral components. Table 1 shows a classification scheme for the phyllosilicates, including layer-lattice clay minerals.[29]

Shale can be defined as an earthy, fine-grained, sedimentary rock with a specific laminated character. Illite appears to be the dominant clay mineral in most of the shales investigated. Chlorite mica is frequently present, smectite is a common component in Mesozoic and younger shales, and kaolinite usually occurs in small amounts only.

According to Rieke and Chilingarian[58] the most important allogenic components of argillaceous sediments are:

(1)   various clay minerals including gibbsite,
(2)   quartz,
(3)   feldspar,
(4)   carbonates,
(5)   amorphous silica and alumina,
(6)   pyroclastic material, and
(7)   organic matter.

Based on a detailed investigation of the analyses of 10 000 shales Yaalon[68] described the mineral composition of the *average shale* as follows:

Clay minerals (predominantly illite)	59 %
Quartz and chert	20 %
Feldspar	8 %
Carbonates	7 %
Iron oxides	3 %
Organic materials	1 %
Others	2 %

† 1 micron = 10^{-4} cm = 10^{-6} m = 1 μm.

TABLE 1

PROPOSED CLASSIFICATION SCHEME FOR THE PHYLLOSILICATES, INCLUDING LAYER-LATTICE CLAY MINERALS[29]

Type	Group (x = layer charge)	Subgroup	Species[a]
	Pyrophyllite–talc x ~ 0	Pyrophyllites	Pyrophyllite
		Talcs	Talc
	Smectite or montmorillonite–saponite x ~ 0·5–1	Dioctahedral smectites or montmorillonites	Montmorillonite, beidellite, nontronite
		Trioctahedral smectites or saponites	Saponite, hectorite, sauconite
	Vermiculite x ~ 1–1·5	Dioctahedral vermiculite	Dioctahedral vermiculite
		Trioctahedral vermiculite	Trioctahedral vermiculite
2:1	Mica[b] x ~ 2	Dioctahedral micas	Muscovite, paragonite
		Trioctahedral micas	Biotite, phlogopite
	Brittle mica x ~ 4	Dioctahedral brittle micas	Margarite
		Trioctahedral brittle micas	Seybertite, xanthophyllite, brandisite
	Chlorite x variable	Dioctahedral chlorites	Penninite, clinochlore, prochlorite
		Trioctahedral chlorites	
2:1:1	Kaolinite–serpentine x ~ 0	Kaolinites	Kaolinite, halloysite
1:1		Serpentines	Chrysotile, lizardite, antigorite

[a] Only a few examples given.
[b] The status of *illite* (or *hydromica*), *sericite*, etc. must at present be left open since it is not clear whether or at what level they would enter the table; many materials so designated may be interstratified.

A typical shale commonly has several components. Hence, no universal shale parameter can be used to characterise a specific type of argillaceous sediment or rock. However, the considerable variations in shale properties are of practical importance in geological concepts, oil-well drilling technology, geophysical formation evaluation, etc.

Density Values of Clay Minerals

Since many clay minerals are subject to isomorphous substitution within the clay lattice, a range of characteristic density values, rather than a single, specific value, best describe their density. Density values typical for several clay minerals have been assembled in Table 2.

TABLE 2

COMPILATION OF CLAY DENSITY DATA

Clay minerals	Density (g/cm^3)	Remarks	References
Chlorite	2·60–2·96	Low water-adsorptive properties	Dana[19]
Halloysite	2·55–2·56	Completely evacuated	Makower[46]
Illite	2·76–3·00	Muscovite	Dana[19]
	2·70–3·10	Biotite	
	2·642–2·688	No absorbed water	DeWit and Arens[21]
Kaolinite	2·609	Theoretical density	Gruner[30]
	2·60–2·68	Extensive literature	Grim[29]
	2·63	Most frequently quoted	
Palygorskite	2·29–2·36	Limited data	Caillere[10]
Sepiolite	2·08	Limited data	Caillere[10]
Smectite	2·20–2·70	Nontronite essentially	Caldwell and Marshall[11]
	2·24–2·30	Saponite dehydrated	
	2·348		DeWit and Arens[21]
	2·2–2·70	Montmorillonite	Grim[29]
	2·53	Low-iron smectite	Makower[46]
	2·74	3·6% iron content	
Vermiculite	No precise data; oven-dried similar to biotite–mica		Grim[29]

Compaction of Argillaceous Sediments

Compaction of argillaceous sediments caused by vertical monotonic loading is a well-established geological phenomenon.[31,32] Factors known to affect the water content of these sediments under applied loads include the type of clay mineral, particle size, absorbed cations, temperature, pH,

Eh, and type of interstitial electrolyte solution.[16] Laboratory compaction experiments illustrate the influence of different factors on the relationship between void ratio and pressure in clayey materials (Fig. 1).

The decrease in porosity is a measure of the extent of compaction an argillaceous sediment has undergone since deposition. Unfortunately, effects due to depositional rates, geological age, subsurface temperature, tectonic forces, etc., modify any simple compaction model.

Most *compaction models* consider clay mineral particles of idealised size and shape, which are affected by mechanical rearrangement during burial. Among the more important compaction models are those proposed by the following workers.

Athy[1]

Compaction causes the squeezing out of formation fluids and this results in a decrease in porosity. In other words, porosity and burial depth of shale are related in an orderly fashion. The degree of compaction, however, is not always *directly* proportional to either the porosity reduction or the bulk density increase. These properties can be affected also by re-crystallisation, solution and cementing effects.

Hedberg[32]

Since various geological and geochemical parameters influence compaction, no single, unique porosity–pressure relationship will characterise all compaction processes. Three distinct compaction stages are proposed:

> Stage I involves dewatering and mechanical rearrangement, the transition from clay to shale, with pressures from $0 \rightarrow 800$ psi, and porosity from $90\% \rightarrow 35\%$;
> Stage II includes mechanical deformation and additional expulsion of absorbed water with pressures from 800 psi $\rightarrow 6000$ psi, and porosity from $35\% \rightarrow 10\%$;
> Stage III is the recrystallisation phase, the transition of shale $\rightarrow$ slate $\rightarrow$ phyllite, with pressures in excess of 6000 psi, and porosity less than 10%.

Weller[67]

Being similar to Hedberg's concept this three-stage compaction model involves:

(1) compacting mud (porosity from $85\% \rightarrow 45\%$),

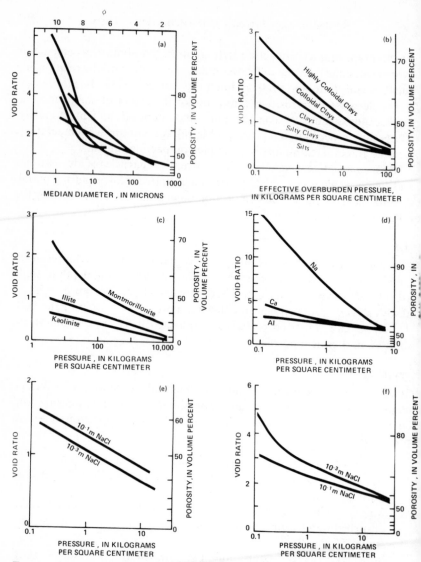

FIG. 1. Influence of different factors on the relationship between void ratio and pressure in clayey materials. (*a*) Relationship between void ratio and median particle diameter at overburden pressures less than 1 kg/cm² (after Meade[51]). (*b*) Generalised influence of particle size (modified from Skempton[61]). (*c*) Influence of clay-mineral species (modified from Chilingar and Knight[12a]). (*d*) Influence of cations absorbed by montmorillonite (modified from Samuels[69]). (*e*) Influence of NaCl concentrations in unfractionated illite, about 60 % of which was coarser than 2 μm in size (modified from Mitchell[51b]). (*f*) Influence of NaCl concentration in illite finer than 0·2 μm (modified from Bolt[5]). (After Meade[51a]); In: Chilingarian and Rieke.[16]

(2)   expulsion of interstitial fluids with concurrent rearrangement of the clay minerals (porosity from $45\% \rightarrow 10\%$), and

(3)   clays squeezed into void space between non-clay minerals, with subsequent deformation and crushing of all grains (porosity from $10\% \rightarrow 0\%$).

Powers[54]

A two-stage water escape theory considers changes in the properties of clay minerals as a function of depth. The model is based on the diagenetic transformation of montmorillonite to illite (Fig. 2). Montmorillonite sediments at a depth of a few hundred feet have reached equilibrium so far as the water in the sediment and the water-absorbing properties of montmorillonite are concerned. Below 1500 ft to 3000 ft a minimum of four mono-molecular water layers (i.e. water of hydration) are stacked between the unit layers of montmorillonite. Below 6000 ft montmorillonite changes to illite and part of the bound water becomes free pore water. At this stage (6000 ft $\rightarrow$ 9000 ft) the resulting particle size decreases, whereas both effective porosity and permeability increase.

Below 9000 ft to 10 000 ft, at the non-montmorillonite level, additional compaction occurs, and both effective porosity and permeability again drastically decrease.

Burst[9]

The two-stage subsurface dehydration model as proposed by Powers previously has been modified by Burst to a three-stage system (Fig. 2). The basic concept is the transformation of montmorillonite to mixed-layer clays, with geothermal effects being considered to be far more important than pressure.

First, mud (density $\approx 1\cdot32$ g/cm^3, porosity $\approx 80\%$) is compacted over the first few thousand feet. After the initial dehydration stage the argillaceous sediments (density $\approx 1\cdot96$ g/cm^3, porosity $\approx 30\%$) continue to absorb geothermal energy (heat). This causes mobilisation of inter-layer water, part of which is discharged into the bulk water system. The amount of water discharged is equivalent to about 10% to 15% of the compacted bulk volume. In addition, mixed-layer clays are formed (density $\approx 2\cdot28$ g/cm^3, porosity $\approx 16\%$).

Water movement during the second stage is capable of re-distributing any mobile subsurface components, such as solutes and hydrocarbons. This large-scale water release will produce fluids at a constant rate under

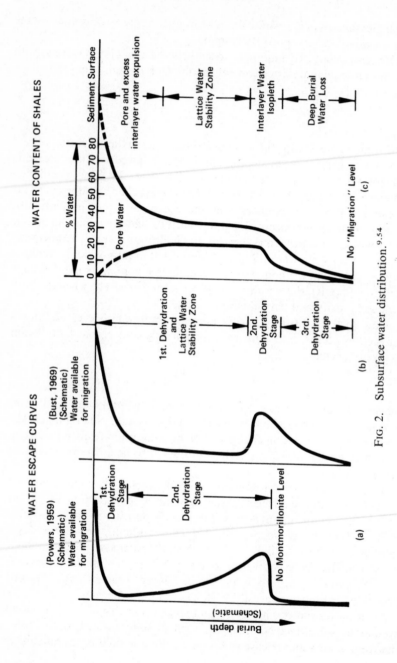

FIG. 2. Subsurface water distribution.[9.54]

geological conditions where the reaction can occur. As such, it should influence any over-pressure environment, either by its mere presence in such a closed system, or by changes imposed on osmotic effects.[25] During the third dehydration stage, which takes place under deep burial and high temperature, additional water loss occurs (density $\approx 2\cdot57\,\text{g/cm}^3$, porosity $\approx 5\%$).

Katz and Ibrahim[40]
This model describes the compaction and fluid expulsion from an argillaceous zone sandwiched between two permeable sand layers. It is based on Terzaghi's spring and piston analogy. According to Chilingarian and Rieke[16] the model explains the higher porosity of under-compacted shales, the extreme drop in permeability with increasing lithostatic pressure, and the occurrence of high interstitial fluid pressure in shales.

Mathematical Models
The complexity of the behaviour of sediments undergoing compaction precludes a rigorous mathematical description of the process. Argillaceous rocks under certain conditions exhibit elastic, visco-elastic or thixotropic properties. Various thermodynamic and geochemical phenomena are also superimposed.

Some of the more promising recent developments in the field of analytical modelling of compaction include the concepts proposed by Gibson,[27] Cooper,[17] Bredehoeft and Hanshaw,[8] Berner,[2] Smith[63] and Rieke and Chilingarian.[58] For example, Chilingarian and Rieke[16] postulated that compaction is directly related to several parameters, and functionally characterised by the following expression:

$$C = \text{f}(\delta, v, \rho, V, \phi, K, D, t, c)$$

where C = compaction, δ = stress on the system, v = velocity term for solids and interstitial fluids in the system, ρ = density, V = volume relationship, ϕ = porosity, K = permeability, D = burial depth, t = time, and c = compressibility relationship.

In addition, the effects of tectonic forces on the compaction mechanism have been investigated in detail.[3,34,35,58,64]

Differential Compaction
Lateral changes in thickness or compactibility of the formations will cause

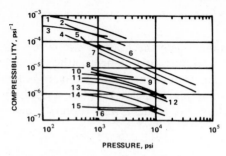

FIG. 3. Relationship between compressibility (psi^{-1}) and applied pressure (psi) for unconsolidated sands, illite clay, limestone, sandstones, and shale.[16]

No.	Investigator	Rock type	Type of applied pressure	Compressibility
1	Chilingarian et al.[15]	California unconsolidated arkosic sands[a]	hydrostatic	pore $[-(1/V_p)(\partial V_p/\partial p_e)_{\bar\sigma}]$
2	Kohlhaas and Miller[42]	California unconsolidated sands	uniaxial	pore
3	Chilingarian et al.[15]	California unconsolidated arkosic sands[a]	hydrostatic	bulk $[-(1/V_b)(\partial V_b/\partial p_e)_{\bar\sigma}]$
4	Kohlhaas and Miller[42]	California unconsolidated sands	uniaxial	bulk
5	Chilingarian et al[15]	Illite clay (API No. 35) (wet)[b]	uniaxial	bulk $[-(1/e + 1)(de/dp_e)]$
6	Chilingarian et al.[15]	Illite clay (API No. 35) (dry)	uniaxial	bulk $[-(1/h)(dh/dp_e)]$
7	Knutson and Bohor[44]	Repetto Formation (Grubb Zone) (wet)[a]	net confining	'pore' $\left.\vphantom{\begin{array}{c}1\\2\\3\end{array}}\right\}[-(1/V_p)(\partial V_p/\partial\bar\sigma)p_p]$
8	Knutson and Bohor[44]	Lansing–Kansas City Limestone (wet)[a]	net confining	'pore'
9	Carpenter and Spencer[12]	Woodbine Sandstone (wet)	net confining	'pseudo-bulk' $[-(1/V_b)(\partial V_p/\partial p)]$
10	Fatt[23]	feldspathic graywacke (No. 10) (wet)[c]	net confining[d]	bulk
11	Fatt[23]	graywacke (No. 7) (wet)[c]	net confining	bulk
12	Fatt[23]	feldspathic graywacke (No. 11) (wet)[c]	net confining	bulk $\left.\vphantom{\begin{array}{c}1\\2\\3\\4\\5\end{array}}\right\}[-(1/V_c)(\partial V_b/\partial p_t)p_p]$
13	Fatt[23]	lithic graywacke (No. 12) (wet)[c]	net confining	bulk
14	Fatt[23]	feldspathic quartzite (No. 20) (wet)[c]	net confining	bulk
15	Podio et al.[53]	Green River Shale (dry)	net confining	bulk
16	Podio et al.[53]	Green River Shale (wet)[b]	net confining	bulk

differential compaction. Conditions which lead to differential compaction include

(1) compactible sediments over erosional unconformities of considerable topographical relief,
(2) sand lenses in massive shales, and
(3) carbonate reefs embedded in shales.

Differential compaction behaviour will also affect specific shale properties, including bulk density, electric resistivity, and acoustic travel time.[24,25]

Compressibility of Argillaceous Sediments

Compressibility is defined as the rate of change of volume per unit volume with respect to the applied stress. Compressibility is important since many properties of clays change with decreasing porosity under high overburden pressure.

The compressibility of clays and shales has been studied by many investigators.[14,15,45,57,60,61,65] Figure 3 shows the relationship between compressibility and applied pressure for samples with differing lithologies. Note the different usages of the term compressibility, depending on the method of measurement.

Tables 3 and 4 summarise compressibility data for several typical clays. Studies by Carpenter and Spencer,[12] Kaul[41] and von Gonten and Choudhary[66] also suggest that rock compressibility increases with increase in temperature.

Conclusion

There exists no unique, universal relationship of any given shale property versus depth which will characterise a specific type of argillaceous sediment. This state of affairs is important with respect to all of the basic concepts involved in applying shale density values in geological studies, oil-well drilling operations, and formation evaluation by means of geophysical well log analyses.

---

[a] Saturated with formation water.
[b] Saturated with distilled water.
[c] Saturated with kerosene.
[d] Net confining pressure $= p_e = (\bar{\sigma} - 0.85 p_p)$, where $\bar{\sigma}$ is the total overburden stress and p_p is the pore pressure.

TABLE 3
VOID RATIO AND COMPRESSIBILITY EQUATIONS FOR VARIOUS CLAYS[a] (FROM RIEKE et al.[57])

Type of clay	Assumed density (g/cm^3)	Relationship between void ratio (e) and effective pressure (p)	Compressibility equations
Montmorillonite[b] (API No. 25)[c]	2·60	$e = 2·69 - 0·467 \, (\log p)$	$c_b = 3·25 \times 10^{-2} p^{-0·874}$
Illite (API No. 35)	2·67	$e = 1·335 - 0·23 \, (\log p)$	$c_b = 3·9 \times 10^{-2} p^{-0·926}$
Halloysite (API No. 12)	2·55	$e = 1·01 - 0·163 \, (\log p)$	$c_b = 3·3 \times 10^{-2} p^{-0·946}$
Kaolinite (API No. 4)	2·63	$e = 0·885 - 0·153 \, (\log p)$	$c_b = 3·5 \times 10^{-2} p^{-0·946}$
Dickite (API No. 15)	2·60	$e = 0·682 - 0·128 \, (\log p)$	$c_b = 3·05 \times 10^{-2} p^{-0·980}$
Hectorite (API No. 34)	2·66	$e = 0·718 - 0·123 \, (\log p)$	$c_b = 2·85 \times 10^{-2} p^{-0·954}$
P-95 dry clay (Buckhorn Lake, California)	2·53	$e = 0·70 - 0·116 \, (\log p)$	$c_b = 2·8 \times 10^{-2} p^{-0·946}$
Soil from limestone terrain (Louisville, Kentucky)	2·67	$e = 0·50 - 0·0816 \, (\log p)$	$c_b = 2·25 \times 10^{-2} p^{-0·952}$

[a] Compressibility equations were calculated from void ratio versus pressure curves $[c_b = -\{1/(e + 1)\}(de/dp)]$.
[b] At pressures above 1000 psi.
[c] American Petroleum Institute, 1951, Preliminary Reports, Reference Clay Minerals, Research Project 49. These clays were obtained from Ward's Natural Science Establishment, Inc., 3000 Ridge Road East, Rochester, New York.

TABLE 4

COMPRESSIBILITY EQUATIONS CALCULATED FROM THICKNESS, h, VERSUS OVERBURDEN PRESSURE, p, MEASUREMENTS FOR VARIOUS CLAYS HYDRATED IN DISTILLED WATER $[c_b = -(1/h)(dh/dp)]$. (REFERENCE 57)

Type of clay	Compressibility equations obtained using different varieties of the same type of clay	
Dickite	$c_b = 3 \cdot 9 \times 10^{-2} p^{-0 \cdot 924}$;	$c_b = 3 \cdot 0 \times 10^{-2} p^{-0 \cdot 788}$
Halloysite	$c_b = 4 \cdot 2 \times 10^{-2} p^{-0 \cdot 848}$;	$c_b = 4 \cdot 7 \times 10^{-2} p^{-0 \cdot 864}$
Hectorite (containing 62% $CaCO_3$)	$c_b = 3 \cdot 4 \times 10^{-2} p^{-0 \cdot 787}$;	$c_b = 4 \cdot 4 \times 10^{-2} p^{-0 \cdot 830}$
Illite	$c_b = 3 \cdot 5 \times 10^{-2} p^{-0 \cdot 815}$;	$c_b = 3 \cdot 7 \times 10^{-2} p^{-0 \cdot 825}$; $c_b = 3 \cdot 8 \times 10^{-2} p^{-0 \cdot 93}$
Kaolinite	$c_b = 4 \cdot 0 \times 10^{-2} p^{-0 \cdot 811}$;	$c_b = 3 \cdot 5 \times 10^{-2} p^{-0 \cdot 784}$
Montmorillonite[a]	$c_b = 2 \cdot 8 \times 10^{-2} p^{-0 \cdot 732}$;	$c_b = 2 \cdot 7 \times 10^{-2} p^{-0 \cdot 735}$

[a] For the best straight line drawn through the experimental points ($p > 1000$ psi). Actually the curve is concave upwards.

SHALE DENSITY MEASUREMENTS

Determination of Shale Bulk Density

Surface Measurements on Subsurface Shale Samples
Bulk density in normally compacted shale increases with depth. Shale bulk density can be determined on drill cuttings, conventional cores, or side-wall core samples retrieved from boreholes. Most frequently it is monitored in drilling by measuring the bulk density of drill cuttings using one of several methods. The latter include:

(1) Fluid density gradient column method;
(2) Float or sink technique, using a series of solutions of different densities;
(3) High-pressure mercury pump method; and
(4) mud balance technique.[4,25,48,50]

Special care must be taken in selecting and preparing the cuttings for analysis. This includes, (a) washing, (b) screening to discard large cavings or smooth recirculated cuttings, and (c) air or spin drying until the cutting surface has a dull appearance. Multiple cuttings have to be tested due to variance in sample data. The average density value at a given depth or interval is then used in the appropriate studies.

Subsurface Measurements Using Geophysical Well-Logging Devices
In geophysical well logging a probe is lowered in the well at the end of an insulated cable, and physical measurements are made and recorded in graphical form as a function of depth. These records are called geophysical well logs, well logs, or simply logs. Often when there is no ambiguity, geophysical well-logging operations are referred to shortly as well logging or logging.

Various types of measuring devices can be lowered on cables in the borehole for the sole purpose of measuring (logging) both borehole and *in situ* formation properties. These logging tools, or logging sondes, contain sensors which measure the desired downhole properties, whether thermal, magnetic, electric, radioactive, or acoustic.

At its inception, the concept of measuring the density of rocks *in situ* was directed more towards obtaining data for gravity studies than for the determination of formation porosity. Today, however, both the borehole gravity meter and the Densilog (trademark of Dresser Atlas) are used for the bulk density measurements necessary for gravity studies, seismic modelling, estimation of reservoir porosity, determination of the elastic constants of rocks, etc.

The borehole gravity meter is basically a density-logging tool. Its radius of investigation, however, is tens of feet, rather than tens of inches as in the case of the Densilog. The borehole gravity meter is a high-precision, large-volume, bulk density tool.[37-39,49,62] The difference in gravitational attraction between any two points below the Earth's surface is a function of the free-air gradient and the mass of material between the two points. This is summarised in the equation:

$$G_1 - G_2 = \Delta G = F\Delta Z - 4\pi K \rho_b \Delta Z$$

where G_1, G_2 = gravitational attraction in milligals at stations 1 and 2, respectively; F = free-air correction, $0 \cdot 09406$ milligal/ft; ρ_b = average bulk density in g/cm^3; K = gravitation constant and ΔZ = difference in elevation in feet. From this:

$$\rho_b(\text{g/cm}^3) = 3 \cdot 687 - 39 \cdot 18 \Delta G / \Delta Z$$

Since ΔG is measured with the down-hole meter and ΔZ with the wireline, down-hole gravity data directly yield bulk density, and no calibration is required. It is also apparent that this bulk density measurement involves a large volume of rock.

The Densilog tool provides a continuous record of variations in the density of the rock interval penetrated by the drill bit. To measure formation bulk density, a beam of gamma rays is directed into the rock. At a fixed distance from the source a counting system detects changes in the intensity of the gamma-ray beam resulting from changes in bulk density of the formation. The higher the density, the lower the intensity of gamma radiation at the detector.[59]

The term bulk density is applied to the overall or gross density of a unit volume of rock. In the case of porous rocks, it includes the density of the fluid in the pore space (i.e. porosity) as well as the grain density of the rock. Formation bulk density (ρ_b) is related to porosity (ϕ), density of fluids (ρ_{fl}) in the porosity, and the matrix grain density (ρ_{ma}) of the rock by:

$$\rho_b = \phi \rho_{fl} + (1 - \phi)\rho_{ma}$$

Hence, formation porosity can be calculated when the bulk density is measured and matrix grain density and fluid density are known:

$$\phi = (\rho_{ma} - \rho_b)/(\rho_{ma} - \rho_{fl})$$

Selection of the proper matrix grain density is based upon knowledge of the lithology of the interval logged.

Factors Influencing Shale Density Values Measured on Drill Cuttings

Besides possible limitations due to (a) cavings and/or recirculated cuttings contaminating the drill cutting samples investigated, and (b) questionable reliability and limited care taken by personnel to catch samples at the shale shaker and in running the tests, several geological factors may greatly affect the measured shale bulk density values. These factors include:

(1) *Shale gas* in the drill cuttings may drastically decrease the apparent bulk density values measured at the surface. Shale gas can be present in the subsurface as a free gas phase or as gas dissolved in the interstitial waters of shale formations. Methane-bearing, under-compacted shales are frequently encountered in worldwide drilling operations. According to Hedberg[33] these shales may be looked upon as petroleum source beds 'caught in the act', i.e. prior to reaching the true petroleum maturation stage.

(2) *Organic rich shales* have bulk density values on the low side.

(3) *Under-compacted, over-pressured shales.* Bulk density in normally compacted shales increases with depth in a regular fashion. Hence, a plot of shale bulk density data versus depth will define the normal compaction trend in any given area. These data can also be used to forecast shale densities at greater depth.

   Since shale porosity increases in pressure transition and over-pressured zones, any decrease in bulk density values may reflect the presence of such abnormal-pressure environments.[25]

(4) *Lithological variations*, such as high carbonate content, the presence of silty or sandy shales, mudstones, marls, etc., may also result in large variations of shale density versus depth.

(5) *Heavy minerals*, such as pyrite (FeS_2), siderite ($FeCO_3$), mica (biotite, muscovite, etc.), may drastically increase the apparent bulk density of shales. For example, pyrite is found in shales in the Permian basin, USA, the South China Sea area, offshore Cameroon in Africa, etc. Mica has been encountered in the North Sea area and siderite has been found in several major petroleum exploration areas around the globe.

(6) *Age boundaries, unconformities, differential compaction*, and *structural* effects, *position* within the clastic basin, etc., may also

greatly affect the normal compaction trend lines, i.e. bulk density versus depth relationship.

However, these factors which markedly influence shale bulk density values are at the same time the foundation for the proper application of such bulk density variations to geological, petrophysical, oil-well drilling, and formation evaluation concepts.

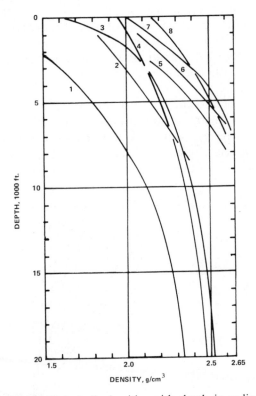

FIG. 4. Variation of shale bulk densities with depth in sedimentary basins. 1 = methane-saturated clastic sedimentary rock—probable minimum density—after McCulloch[49]; 2 = mudstone, Po Valley Basin, Italy (after Storer[61a]); 3 = average Gulf Coast shale densities derived from density logs and formation samples (after Eaton[22]); 5 = Motatan-1, Maracaibo Basin, Venezuela (after Dallmus[18]); 6 = Gorgeteg No. 1 Hungary, calculated wet density values (after Skeels[59a]); 7 = Pennsylvanian and Permian dry shale values, Oklahoma and Texas; Athy's adjusted curve (after Dallmus[18]); 8 = Las Ollas-1, Eastern Venezuela (after Dallmus[18]). In Chilingar and Rieke.[13]

APPLICATION TO GEOLOGICAL CONCEPTS

Over the first few hundred feet of burial, during gravitational compaction of argillaceous sediments, the bulk density increases rapidly with depth. At great depth the changes are small and the bulk density values approach the weighted average of the grain densities. Basically, the incremental increase in bulk density with depth is a function of numerous factors. These include overburden and tectonic stresses, geological time, subsurface temperature, mineral components (type of clays, impurities), grain-size distribution, etc. The observed change of shale bulk densities with depth in various sedimentary basins is shown in Fig. 4.

The mathematical relationship between shale porosity and the density with increasing geostatic loading (i.e. depth) as proposed by Ozerskaya[52] is presented in Fig. 5.

A recent study of Miocene shales in California has shown a considerable, yet not systematic, lateral variation in bulk density.[20] Dallmus,[18] however, showed a systematic increase of bulk density in shale formations from the rim towards the centre of a sedimentary basin in South America. This is shown schematically in Fig. 6.

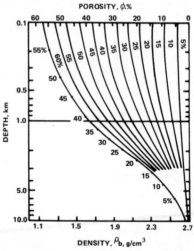

FIG. 5. Relationship between bulk density, porosity and depth of burial (geostatic load) for argillaceous sediments. $\phi = \phi_{max}\exp(-0.45D)$; $\rho_b = \rho_s(1 - \phi_{max}\exp(-0.45D))$; $\rho_s = 2.7$. The numbers designate the values of initial porosity ($\phi = \phi_{max}$) from 60% to 5% (with 5% intervals). The same numbers shown on the first curve to the left, correspond to curves with different initial porosity, shifted along the depth scale to the 60% curve.[52]

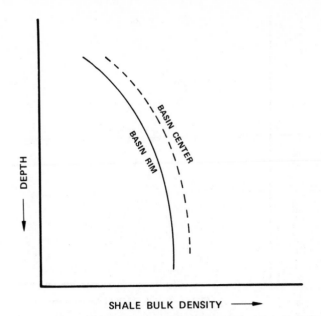

FIG. 6. Relationship of shale bulk density versus depth as a function of location within basin (rim versus centre.)

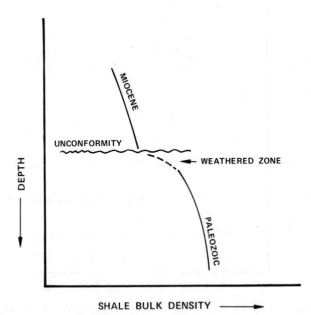

FIG. 7. Relationship of shale bulk density versus depth across an unconformity.

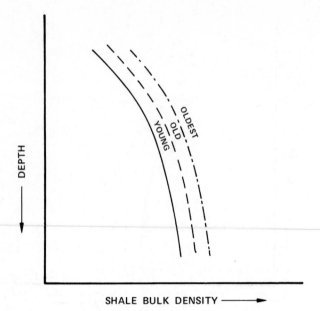

FIG. 8. Relationship of shale bulk density versus depth as a function of geological age.

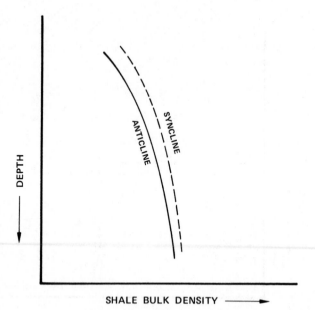

FIG. 9. Relationship of shale bulk density versus depth as a function of anticline versus syncline.

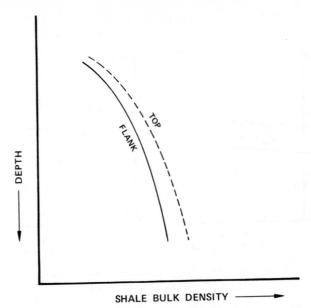

FIG. 10. Relationship of shale bulk density versus depth as a function of structural position (top versus flank–differential compaction).

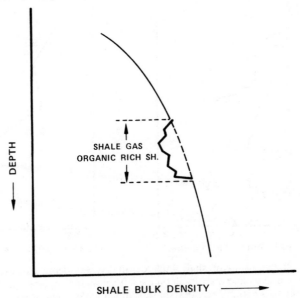

FIG. 11. Relationship of shale bulk density versus depth as affected by shale gas and/or organic-rich shales.

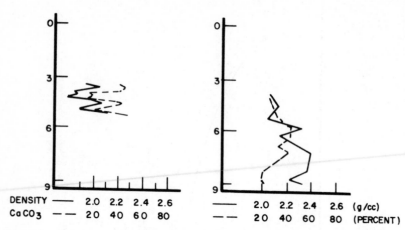

FIG. 12. Relationship of shale bulk density versus depth. Effect of the carbonate content on the density of shales.

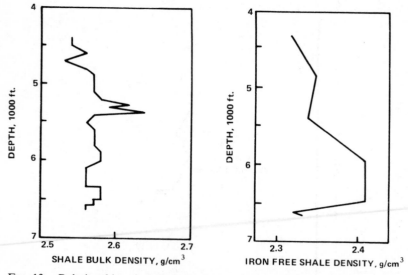

FIG. 13. Relationship of shale bulk density versus depth as affected by the presence of heavy minerals. Graph on the left shows the uncorrected shale density plotted versus depth. Since the shale samples contained siderite ($FeCO_3$) in concentrations varying up to 7 %, the bulk density, corrected for this siderite effect, is replotted versus depth in the graph on the right.

Marked bulk-density variations in shales with depth assist in detecting major unconformities. A typical variation in bulk density across an unconformity is illustrated in Fig. 7. Note the indication of the presence and thickness of the weathered zone right at the unconformity.

A systematic increase in shale bulk density may be also observed with increasing geological age (Fig. 8) or as a function of shale density variations across an anticline versus syncline (Fig. 9).[18] Differential compaction effects are shown in Fig. 10.

The presence of organic-rich shales or shale gas can be responsible for a marked decrease in bulk density (Fig. 11), whereas calcareous material (Fig. 12) and the presence of heavy minerals (pyrite, siderite, mica, etc.) will also greatly increase the measured bulk density values (Fig. 13).

The use of shale bulk density data to detect the presence and estimate the magnitude of abnormally high formation pore pressures is a valuable aid in petroleum exploration and drilling operations (Fig. 14). This will be discussed next, in more detail.

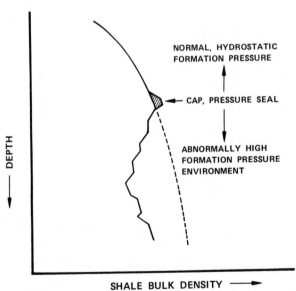

FIG. 14. Relationship of shale bulk density versus depth in normal (i.e. hydrostatic) and abnormally high formation pressure environments. Note the presence of the calcareous barrier (i.e. cap, pressure seal) separating hydrostatic pressures and overpressures. In abnormally high formation pressures the shale bulk density shows departure, i.e. decrease, from the normal compaction trend. Amount of departure can be related to magnitude of overpressures.

APPLICATION TO OIL-WELL DRILLING OPERATIONS

Basic Concepts of Detection and Evaluation of Abnormal Formation Pressure Environments Using Shale Bulk Density Data

Hydrostatic pressure is determined by the unit weight and vertical height of a fluid column. The hydrostatic pressure gradient is affected by the concentration of dissolved solids and gases in the fluid column and variations in the temperature gradient. For example, a pressure gradient of 0·465 psi/ft (0·1074 kgcm^{-2}m^{-1}) assumes a water salinity of 80 000 ppm sodium chloride at a temperature at 77 °F (25 °C).

Overburden pressure originates from the combined weight of the formation matrix and the fluids (water, oil and gas) in the pore space overlying the formation of interest. For example, average Tertiary deposits on the US Gulf Coast exert an overburden pressure gradient of about 1·0 psi/ft of depth (0·231 kgcm^{-2}m^{-1}). This corresponds to a force exerted by a formation with an average bulk density of 2·31 g/cm^3.

Formation pressure is the pressure acting upon the fluids (water, oil and gas) in the pore space of the formation. Normal formation pressures in any geological setting will equal the hydrostatic head (i.e. hydrostatic pressure) of water from the surface to the subsurface formation. Formation pressures exceeding hydrostatic pressures in a specific geological environment are defined as abnormally high formation pressures (surpressures), whereas formation pressures less than hydrostatic are called subnormal (subpressures).

The ability to locate and quantitatively evaluate over-pressured formations, which occur worldwide in the Earth's crust, is critical in drilling operations, completion techniques, and the development of exploratory and reservoir engineering concepts for a successful search and optimum exploitation of the world's valuable, but exhaustible, oil and gas resources. The importance of this to the oil industry has recently been summarised.[25]

Shale bulk density values are one of the most useful pressure indicator methods. Basically, shale bulk density values are plotted versus depth and normal trend lines are established. Since shale porosity increases in pressure transition and overpressured zones, any decrease in bulk density values may reflect the presence and magnitude of such abnormal-pressure environments (Fig. 15).

Quantitative pressure evaluation from shale density data is possible by the equivalent depth method or from empirical curves established for a given area. Proposed modifications of these basic concepts will yield only minor variations in the quantitative analysis.[28]

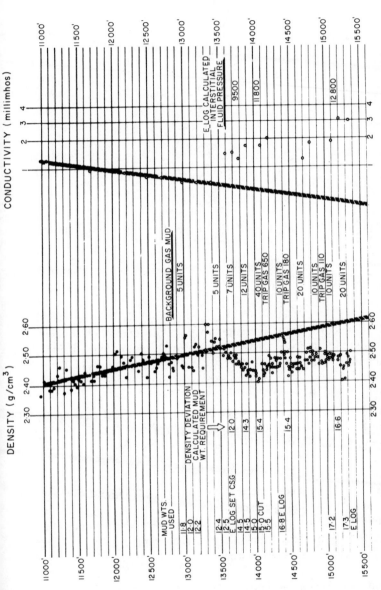

Fig. 15. Shale bulk density variations in normal and overpressured zones plus well-log derived conductivity curve and drilling mud weight requirements.[4]

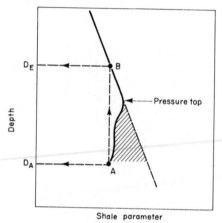

Shale parameter

FIG. 16.   Schematic representation of D_A and D_E using the equivalent depth method.

Quantitative pressure evaluation can be made using the equivalent depth method, in which:

$$P_f = G_o D_A - D_E(G_o - G_H)$$

where P_f = pore fluid pressure (psi), D_A = depth of interest in overpressured interval (ft), D_E = normal equivalent depth (ft) corresponding to D_A (see Fig. 16), G_H = hydrostatic gradient in the subject well (psi/ft), and G_0 = overall overburden gradient in the subject well (psi/ft).

For example, in areas where the overburden gradient equals $1 \cdot 0$ psi/ft $(2 \cdot 31 \, \text{kgcm}^{-2} \text{m}^{-1})$ and the hydrostatic gradient equals $0 \cdot 465$ psi/ft $(0 \cdot 107 \, \text{kgcm}^{-2} \text{m}^{-1})$, this relationship can be simplified to:

$$P_f = D_A - 0 \cdot 535 D_E$$

This relationship has been successfully used in the US Gulf Coast area, the North Sea region, and elsewhere.

Despite shortcomings, such as the effects due to the presence of a high carbonate content, heavy minerals, etc., shale bulk density plots are very helpful in pressure evaluation work.

Application in Formation Fracture Pressure Gradient Studies

Maximum well control and minimum cost are key factors in present-day drilling and completion practices. To avoid, or at least minimise, the danger of well kicks, stuck pipe, and lost circulation, a basic understanding of formation pore pressure and fracture pressure is necessary. Pore pressure and fracture gradient information are the two basic parameters for any well drilling and completion plan; they are equally important in cementing, hydraulic fracturing, fluid injection, sand consolidation, etc.

Generally speaking, formation fracture pressure gradients are related to formation pore pressure, lithology, age and depth of the formation, and to the *in situ* rock stress level.

Drilling mud weights exceeding pore pressures by about 0·2 to 0·4 ppg (pounds per gallon) allow the drill string to be pulled without swabbing, result in reasonable mud costs, good penetration rates, and they reduce the possibility of differential pipe sticking. On the other hand, wells tend to kick when the actual mud weight in the borehole is lower than is needed to control excessively high pore pressure in any permeable, yet uncased, interval in the section penetrated by the drill bit.

Differential pipe sticking may be due to such under-balance (negative differential sticking), but may also be caused by large over-balance (positive differential sticking). In most circumstances, the prevention of lost circulation depends on the utilisation of reliable estimates of fracture pressure gradients and pore pressures. Often large over-balance causes the protection pipe to be set far too high to cure lost circulation problems. These concepts and the associated problems have been reviewed in detail recently.[25]

Various methods have been developed and are employed in the oil industry to calculate hydraulic fracture pressure gradients. One of the best techniques available today utilises shale bulk density and, hence, will be discussed next in some detail.

Method of Eaton
Assuming Earth formations to be elastic, Eaton[22] related the horizontal stress, σ_H, and the vertical stress. σ_v. by Poisson's ratio, v, as expressed in Hooke's law:

$$\sigma_H = \frac{v\sigma_v}{(1 - v)}$$

Eaton then extended the concept proposed earlier by Matthews and Kelly[47] by introducing Poisson's ratio into the expression for the fracture pressure gradient (FPG):

$$FPG = P_f/D + \frac{v}{1 - v}\sigma/D$$

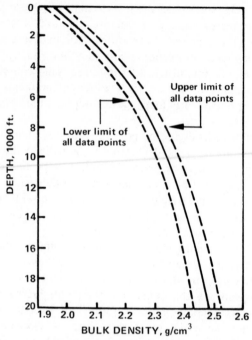

FIG. 17. Shale bulk density variations with depth in US Gulf Coast shales.[22]

where σ = matrix stress which equals $(P_o - P_f)$, P_o = overburden pressure (psi), P_f = formation pore pressure (psi), and D = depth (ft). Basically, the author assumed all of the independent variables to be functions of depth, i.e. overburden load and Poisson's ratio.

Generally speaking, the overburden pressure at a given depth D equals the cumulative weight of the rocks above that depth:

$$P_o = 0.4335 \int_0^D \rho_b(h) \, dh$$

where $\rho_b(h)$ = formation bulk density as a function of depth h, and 0.4335 = a constant for converting g/cm^3 to psi/ft. Hence, local or regional overburden pressures can be determined by using a simple plot of shale bulk density versus depth (derived from density logs or other shale density measurements) and converting them into an overburden load gradient curve. Figures 17 and 18 illustrate this procedure for typical US Gulf Coast conditions as derived by Eaton.[22]

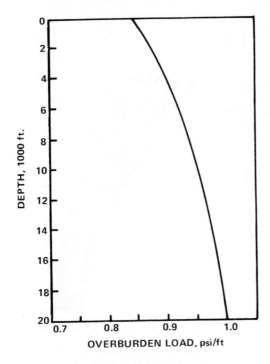

FIG. 18. Relationship between the overburden load gradient and depth for US Gulf Coast shales, which exhibit the bulk density versus depth behaviour as illustrated in Fig. 17.[22]

Besides overburden-load gradient and fracture-pressure gradient data at several depths (such as actual fracturing data, cement squeeze, and lost circulation values), knowledge of the formation pore pressures at corresponding depths is required. Using these data and rewriting the above relationship as:

$$\frac{v}{1 - v} = \frac{\text{FPG} - P_f/D}{(\sigma/D)}$$

the Poisson's ratios in a given area can be back-calculated and plotted versus depth. Simplified local or regional fracture-pressure gradient prediction charts can then be developed, which assist in any engineered well planning. Figure 19 shows such a chart for the US Gulf Coast area.

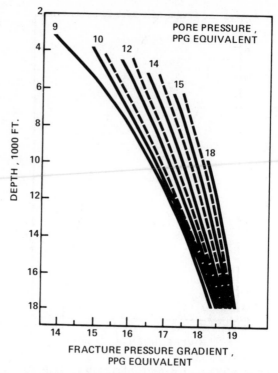

FIG. 19. Nomograph for graphical fracture-pressure gradient estimates for the US Gulf Coast. Based on the model proposed by Eaton.[22] (PPG = pounds per gallon.)

APPLICATION TO FORMATION EVALUATION USING GEOPHYSICAL WELL LOGS

The Densilog responds to electron density, but usually the electron density is closely proportional to the bulk density of the formation. The borehole effect is very small and the mud-cake effect is automatically allowed for in the newer tools. Furthermore, density-derived porosities require no correction for lack of compaction. However, since the tool has a shallow depth of investigation, it is affected by the presence of residual hydrocarbons. The readings are recorded on a linear scale. It is of utmost importance that the side of the tool maintains good contact with the wall of the hole or the apparent density reading will be too low. Normally, permeable beds do not cave appreciably, so that reliable readings are usually available for the zones of interest.

In water-bearing, clean sandstones the Densilog measures the effective porosity (ϕ_e) directly, so that:

$$\phi_D = \frac{\rho_{ma} - \rho_b}{\rho_{ma} - \rho_{fl}} = \phi_e$$

where ρ_b = formation bulk density from log, ρ_{ma} = grain density (US Gulf Coast, 2·65–2·68 g/cm³), ρ_{fl} = pore fluid density: fresh mud filtrate is 1·0 g/cm³; for salt muds, the fluid density becomes 1·15 g/cm³, and ϕ_e = effective porosity of the formation, i.e. only the fluid-filled fraction of the formation.

In water-bearing, shaly formations and considering a dispersed clay system, the shale effect is taken into account by:

$$\phi_e = \frac{(\rho_{ma} - \rho_b)(1 - q)}{\rho_{ma} - \rho_{fl} - q(\rho_{sh} - \rho_{fl})}$$

where q = fraction of total porosity occupied by dispersed clay, and ρ_{sh} = density of dispersed clay minerals (many US Gulf Coast shaly sands have $\rho_{ma} = \rho_{sh}$).

Another procedure for expressing the shale effect is:

$$\phi_D = \phi_e + V_{sh} \cdot \frac{\rho_{ma} - \rho_{sh}}{\rho_{ma} - \rho_{fl}}$$

where V_{sh} = fraction of bulk volume. Since ρ_{sh} usually differs little from ρ_{ma}, the coefficient of V_{sh} ranges in most cases from 0 to 0·2.

In hydrocarbon-bearing formations the Densilog is affected by the presence of hydrocarbons near the borehole. Light oil and especially gas, because of its low density, have the most pronounced effect. The measured bulk density will be too low and thus the apparent porosity too high. Mathematical relationships allow the proper corrections to be made.[26]

In shales the Densilog will measure the bulk density directly. However, severe borehole enlargements (washouts), the presence of shale gas, heavy minerals, etc., will affect the logging measurements. This has to be considered if such logging data are applied in geological studies, synthetic seismograms, etc.

The borehole gravity meter allows highly sensitive measurements of the vertical gravitational attraction at points within the borehole. Hence, an estimate of the bulk density in the interval between two successive gravity measurements can be obtained. Since these data are relatively insensitive to

TABLE 5
DEFINITION OF ELASTIC CONSTANTS.[43]

Elastic constants	Basic equations	Interrelation of equations	Equations in well logging terms
Young's Modulus [a]	$E = \dfrac{9K\rho V_s^2}{3K + \mu V_s^2}$	$E = \dfrac{3K\mu}{3K+\mu} = 2\mu(1+\sigma) = 3K(1-2\sigma)$	$E = \left(\dfrac{\rho}{\Delta t_s^2}\right)\left(\dfrac{3\Delta t_s^2 - 4\Delta t_c^2}{\Delta t_s^2 - \Delta t_c^2}\right) \times 1\cdot34 \times 10^{10}$
Bulk Modulus [b]	$K = \rho\left(V_c^2 - \tfrac{4}{3}V_s^2\right)$	$K = \dfrac{E\mu}{3(3\mu - E)} = \mu\dfrac{2(1+\sigma)}{3(1-2\sigma)} = \dfrac{E}{3(1-2\sigma)}$	$K = \rho\left(\dfrac{3\Delta t_s^2 - 4\Delta t_c^2}{3\Delta t_s^2 \Delta t_c^2}\right) \times 1\cdot34 \times 10^{10}$
Shear Modulus [c]	$\mu = \rho V_s^2$	$\mu = \dfrac{3KE}{9K - E} = 3K\dfrac{1 - 2\sigma}{2 + 2\sigma} = \dfrac{E}{2+2\sigma}$	$\mu = \left(\dfrac{\rho}{\Delta t_s^2}\right) \times 1\cdot34 \times 10^{10}$
Poisson's ratio [d]	$\sigma = \dfrac{1}{2}\dfrac{\left(\dfrac{V_c^2}{V_s^2}\right) - 2}{\left(\dfrac{V_c^2}{V_s^2}\right) - 1}$	$\sigma = \dfrac{3K - 2\mu}{2(3K+\mu)} = \left(\dfrac{E}{2\mu} - 1\right) = \dfrac{3K - E}{6K}$	$\sigma = \dfrac{1}{2}\left(\dfrac{\Delta t_s^2 - 2\Delta t_c^2}{\Delta t_s^2 - \Delta t_c^2}\right)$

ρ = bulk density (g/cm³)
V_s = shear velocity (ft/s)
Δt_s = shear travel time (μs/ft)

V_c = compressional velocity (ft/s)
Δt_c = compressional travel time (μs/ft)
$1\cdot34 \times 10^{10}$ = conversion factor

[a] Young's modulus (E) measures opposition of a substance to extensional stress, $E = \dfrac{F/A}{\Delta l/l}$.

[b] Bulk modulus (K) is the coefficient of incompressibility and measures opposition of a substance to compressional stress, $K = \dfrac{F/A}{\Delta r/r}$.

[c] Shear modulus (μ), also called rigidity modulus, measures the opposition of a substance to shear stresses. Finite values for solids, zero values for fluid, $\mu = \dfrac{F/A}{\tan S}$.

[d] Poisson's ratio (σ) is the ratio of relative decrease in diameter to relative elongation, $\sigma = \dfrac{\Delta d/d}{\Delta l/l}$.

borehole geometry, a borehole gravity survey provides very precise estimates of bulk density, particularly in poorly consolidated sediments and in cased well bores.

Recent technical literature,[6,7,36,55,56] also illustrates the application of the borehole gravity meter (of the LaCoste and Romberg type) for sensing density variations which are not actually intersected by the borehole. However, independent density information, such as from cores or the Densilog, is a pre-requisite for this type of application.

Knowledge of bulk density and acoustic well logging data allows the continuous determination of the four basic elastic rock properties, such as Young's modulus, bulk modulus, shear modulus, and Poisson's ratio, over the entire length of the borehole. Table 5 summarises the definitions and mathematical relationships for these elastic constants.[43]

The application of these calculated elastic constants is manifold. It includes various aspects of rock mechanics, soil and foundation studies, dam-site selection, roof and floor rock properties relating to mining operations, drilling and stimulation of oil and gas wells, formation consolidation characteristics, prediction of sand production in high-yield oil and water wells, etc.

REFERENCES

1. ATHY, L. F. (1930). *Am. Assoc. Petrol. Geologists Bull.*, **14**(1), 1–24.
2. BERNER, R. A. (1971). *Principles of Chemical Sedimentology*, McGraw-Hill, New York.
3. BERRY, F. A. F. (1969). *J. Petrol. Tech.*, **21**, 13–14.
4. BOATMAN, W. A. (1967). *World Oil*, **165**(2), 69–74. Also in *J. Petrol. Tech.*, **19**, 1967, 1423–31.
5. BOLT, G. H. (1956). *Geotechnique*, **6**(2), 86–93.
6. BRADLEY, J. W. (1974). *Symp. Houston Geophys. Soc.*, Technical Paper, Houston, Texas.
7. BRADLEY, J. W. (1975). *45th Ann. Meet. Soc. Explor. Geophysicists*, Technical Paper, Denver, Colorado.
8. BREDEHOEFT, J. D. and HANSHAW, B. B. (1968). *Geol. Soc. Am. Bull.*, **79**, 1097–1106.
9. BURST, J. F. (1969). *Am. Assoc. Petrol. Geologists Bull.*, **53**, 73–93.
10. CAILLERE, S. (1948). *Compt. Rend.*, **227**, 855–6.
11. CALDWELL, O. G. and MARSHALL, R. (1942). *Coll. Agr. Res. Bull.*, 354.
12. CARPENTER, C. and SPENCER,, G. B. (1940). *US Bur. Mines*, Rep. Invest. No. 3540.
12a. CHILINGAR, G. V. and KNIGHT, L. (1960). *Am. Assoc. Petrol. Geologists Bull.*, **44**, 101–6.

13. CHILINGAR, G. V. and RIEKE, H. H. (1974). *Compaction of Argillaceous Sediments*, Elsevier, Amsterdam.
14. CHILINGAR, G. V., RIEKE, H. H. and SAWABINI, C. T. (1969). *Symp. Land Subsidence*, Int. Assoc. Sci. Hydrol. and UNESCO, Tokyo, Japan, **89**(2), 377–93.
15. CHILINGARIAN, G. V., SAWABINI, C. T. and RIEKE, H. H. (1973). *J. Sediment. Petrol.*, **43**, 529–36.
16. CHILINGARIAN, G. V. and RIEKE, H. H. (1976). Chapter II, in: *Abnormal Formation Pressures—Implications to Exploration, Drilling and Production of Oil and Gas Resources* (W. H. Fertl, ed.), Elsevier, Amsterdam.
17. COOPER, H. H. (1966). *J. Geophys. Res.*, **71**, 4785–90.
18. DALLMUS, K. F. (1958). In: *Habitat of Oil* (L. G. Weeks, ed.), Am. Assoc. Petrol. Geologists Mem., 36, 2071–2124.
19. DANA, E. S. (1921). *Textbook of Mineralogy* (W. E. Ford, ed.), Wiley, New York.
20. DANA, S. W. (1967). *Compass*, **44**(4), 172–8.
21. DEWIT, C. P. and ARENS, B. (1950). *Trans. 4th Internat. Congr. Soil Sci.*, **2**, 59–62.
22. EATON, B. A. (1969). *J. Petrol. Technol.*, **21**, 1353–60.
23. FATT, I. (1958). *Am. Assoc. Petrol. Geologists Bull.*, **42**, 1924–57.
24. FERTL, W. H., PILKINGTON, P. E. and REYNOLDS, E. B. (1975). *Petroleum Engineer*, **47**(4), 40–7.
25. FERTL, W. H. (ed.) (1976). *Abnormal Formation Pressures—Implications to Exploration, Drilling, and Production of Oil and Gas Resources*, Elsevier, Amsterdam.
26. GAYMARD, R. and POUPON, A. (1967). *Response of Neutron and Formation Density Logs in Hydrocarbon-bearing Formations*, Société de Prospection Electr., Schlumberger, Paris, France.
27. GIBSON, R. E. (1958). *Geotechnique*, **8**, 171–82.
28. GRIFFIN, D. G., BAZER, D. A. (1968). SPE 2166, *AIME Fall Meet.*, Houston, Texas.
29. GRIM R. E. (1968). *Clay Mineralogy*, Mc-Graw-Hill, New York.
30. GRUNER, J. W. (1937). *Am. Mineralogist*, **22**, 855–60.
31. HEDBERG, H. D. (1926). *Am. Assoc. Petrol. Geologists Bull.*, **10**, 1035–72.
32. HEDBERG, H. D. (1936). *Am. J. Sci.*, **31**, 241–87.
33. HEDBERG, H. D. (1974). *Am. Assoc. Petrol. Geologists Bull.*, **58**, 661–73.
34. HERGERT, F. (1973). *Int. J. Rock Mech. Min. Sci.*, **10**, 37–51.
35. HUBBERT, M. K. and RUBEY, W. W. (1959). *Geol. Soc. Am. Bull.*, **70**, 115–66.
36. JAPELER, A. H. (1975). *50th Ann. Meet. Soc. SPE of AIME*, Technical Paper, Dallas, Texas.
37. JONES, B. R., JAGELER, A. H. and NETTLETON, L. L. (1971). *41st Ann. SEG Meeting*, Technical Paper, Houston, Texas.
38. JONES, B. R. (1972). *13th Prof. Well Log Analysts Symp.*, Technical Paper, Tulsa, Oklahoma.
39. JONES, B. R. (1972). *World Oil*, **175**(2), 56–9.
40. KATZ, D. L. and IBRAHIM, M. A. (1971). SPE 322, *5th Conf. on Drilling and Rock Mechanics*, Austin, Texas.

41. KAUL, B. K. (1965). Behaviour of soils under stress, *Indian Inst. Sci. Proc.*, 1(A8), 31.
42. KOHLHAAS, C. A. and MILLER, F. G. (1969). SPE 2563, *AIME Fall Meet.*, Denver, Colorado.
43. KOWALSKI, J. and FERTL, W. H. (1976). *Erdol-Erdgas Z.*, **92**, 301–5.
44. KNUTSON, C. F. and BOHOR, B. F. (1963). in: *Rock Mechanics* (C. Fairhurst, ed.), Pergamon, New York.
45. MACEY, H. H. (1940). *Proc. Phys. Soc. (Lond.)*, **52**(5): 625–56.
46. MAKOWER, B. (1937). *Soil Sci. Soc. Am. Proc.*, **2**, 101–9.
47. MATTHEWS, W. R. and KELLY, J. (1967). *Oil and Gas J.*, **65**(8), 92–106.
48. MATTHEWS, W. R. (1969). *Oil and Gas J.*, **67**(44), 69–76.
49. MCCULLOCH, T. H. (1967). *US Geol. Surv.*, Prof. Paper, 528A.
50. MCKEE, R. E. and PILKINGTON, P. E. (1964). *Oil and Gas J.*, Dec. 2, 102–4.
51. MEADE, R. H. (1964). *US Geol. Surv.*, Prof. Paper 497B.
51a. MEADE, R. H. (1968). *US Geol. Surv.*, Prof. Paper, 497D, 1–39.
51b. MITCHELL, J. K. (1960). In: *Interparticle Forces in Clay–Water–Electrolyte Systems*. (R. H. G. Parry, ed.), Commonwealth Sci. Ind. Res. Organization, Melbourne, 2-92–2-97.
52. OZERSKAYA, M. L. (1965). *Izv. Akad. Nauk SSSR; Ser. Phys. Earth*, 1965(1), 103–8.
53. PODIO, A. L., GREGORY, A. R. and GRAY, K. E. (1968). *J. Soc. Pet. Eng.*, **8**, 389–404.
54. POWERS, M. C. (1967). *Am. Assoc. Petrol. Geologists Bull.*, **51**(7), 1240–54.
55. RASMUSSEN, N. F. (1973). *14th Prof. Well Log Analysts Symp.*, Technical Paper, McAllen, Texas.
56. RASMUSSEN, N. F. (1975). *Log Analyst*, **5**, 3–10.
57. RIEKE, H. H., GHOSE, S. K., FAHHAD, S. A. and CHILINGAR, G. V. (1969). *Proc. Int. Clay Conf.*, **1**, 817–28.
58. RIEKE, H. H. and CHILINGARIAN, G. V. (1974). *Compaction of Argillaceous Sediments*, Elsevier, Amsterdam.
59. SCHNEIDER, L. E. and WATT, H. B. (1973). 'Densilog', Technical Bulletin, Dresser Atlas, Houston, Texas.
59a. SKEELS, D. C. (1943). Standard Oil Co. (N.J.), Geophys. Res. Bull., No. 2, 14 pp.
60. SKEMPTON, A. W. (1945). *Quart. J. Geol. Soc. Lond.*, **100**, 119–35.
61. SKEMPTON, A. W. (1953). *Proc. Yorks Geol. Soc.*, **29**, 33–62.
61a. STORER, D. (1959). In: *I Giacementi Gassiferi dell' Europa Occidentale*, **2**. Acad. Nazl. dei Lincei, Roma, 519–36.
62. SMITH, N. J. (1950). *Geophysics*, **15**, 605–35.
63. SMITH, J. E. (1971). *Math. Geol.*, **3**, 239–63.
64. THOMPSON, T. L. (1973). *Am. Assoc. Petrol. Geologists Bull.*, **57**, 1844–5.
65. VAN DER KNAAP, W. and VAN DER VLIS, A. C. (1967). *Proc. 7th World Petroleum Congress*, Mexico City, **3**, 85–95.
66. VON GONTEN, W. D. and CHOUDHARY, B. K. (1969). SPE 2526, 44th Ann. *AIME Meet.*, Denver, Colorado.
67. WELLER, J. M. (1959). *Am. Assoc. Petrol. Geologists Bull.*, **43**(2), 273–310.
68. YAALON, D. H. (1962). *Clay Minerals Bull.*, **5**(27), 31–6.
69. SAMUELS, S. G. (1950). Building Research Station (Great Britain), Note C176, 16 pp.

INDEX